AN INTRODUCTION TO ISOZYME TECHNIQUES

An Introduction to Isozyme Techniques

George J. Brewer

Department of Human Genetics
University of Michigan Medical School
Ann Arbor, Michigan

With a contribution by

Charles F. Sing

Department of Human Genetics
University of Michigan Medical School
Ann Arbor, Michigan

Academic Press

New York and London

1970

COPYRIGHT © 1970, BY ACADEMIC PRESS, INC.
ALL RIGHTS RESERVED
NO PART OF THIS BOOK MAY BE REPRODUCED IN ANY FORM,
BY PHOTOSTAT, MICROFILM, RETRIEVAL SYSTEM, OR ANY
OTHER MEANS, WITHOUT WRITTEN PERMISSION FROM
THE PUBLISHERS.

ACADEMIC PRESS, INC.
111 Fifth Avenue, New York, New York 10003

United Kingdom Edition published by
ACADEMIC PRESS, INC. (LONDON) LTD.
Berkeley Square House, London W1X 6BA

LIBRARY OF CONGRESS CATALOG CARD NUMBER: 78-116700

PRINTED IN THE UNITED STATES OF AMERICA

To My Mother and Father

Contents

Foreword, ix
Preface, xi

1. Introduction

    A. Strengths of the Isozyme Approach, 1
    B. History, 2
    C. Principles, 6
    D. Terminology and Definitions, 8
    References, 14

2. Starch Gel Electrophoresis

    A. Introduction, 16
    B. Starch Gel Trays and Related Parts, 17
    C. Preparation of the Starch Gel, 21
    D. Adding the Sample, 23
    E. Electrophoretic Setup, 24
    F. Electrophoresis, 30
    G. Slicing the Gel, 31
    H. Staining the Gel, 34
    I. Gel Preservation, 37
    J. Records, 37
    K. Glycerolization of Gels, 38
    L. Destaining of Gels, 38
    M. Special Studies, 38
    References, 39

3. Electrophoretic Media Other than Starch Gel

    A. Introduction, 40
    B. Paper Electrophoresis, 40
    C. Agar Gel Electrophoresis, 43
    D. Cellulose Acetate Electrophoresis, 45
    E. Acrylamide Gel Electrophoresis, 47
    F. Choice of Media, 51
    References, 52

4. Sample Selection and Preparation

    A. Introduction, 53
    B. Specific Procedures, 53
    References, 61

5. Specific Electrophoretic Systems

    A. Introduction, 62
    B. General Comments about Buffer Systems, 63
    C. Staining Systems, 63
    D. Reagents, 67
    E. Isozyme Methods Employed in Our Laboratory, 70
    F. Other Methods, 132
       References, 134

6. Present Applications and the Future of Isozymology

    A. Introduction, 138
    B. Clinical Applications, 138
    C. Somatic Cell Genetics, 143
    D. Tissue-Organ and Intracellular Differentiation, 145
    E. Developmental Genetics, 146
    F. Genetic Variation, 147
    G. Substrate Specificities and Biochemical Relationship of Enzymes, 151
    H. Plant Isozymes, 152
    I. The Use of Isozymes for the Study of Evolution, 153
       References, 157

7. Analysis of Electrophoretic Variation
(*Coauthored with Charles F. Sing*)

    A. Introduction, 158
    B. Data Acquisition, 159
    C. Data Reduction, 168
       References, 175

Author Index, 177
Subject Index, 182

Foreword

The fact that different molecular forms of enzymes which catalyze the same reaction (isozymes) can occur in the same organism is proving to be not only an especially valuable aid in many biological studies, but it is also providing a new and exciting perspective for the interpretation of a number of problems central to modern biological thought, such as cellular differentiation, ontogenetic development, and evolution.

Although the concept of isozymes has been known for some time, the possibility that the different enzyme forms are not produced by *in vitro* effects brought about by preparative procedures but can actually represent distinct molecules has been clarified only within the past decade. Most of the initial evidence was based on immunochemical tests and on the finding of inherited variants of isozymes, which strongly indicated that some of the different enzymes comprising an isozymic group were coded by separate genes. Thus, well in advance of the complete characterization of the primary structure of individual isozymes, the genetic and immunochemical evidence clearly anticipated the molecular basis of isozymes. However, in spite of this convincing evidence, it is somewhat surprising that many biologists and biochemists at that time did not appreciate the importance of isozymes. Much of the early work was carried out by a few biochemically oriented biologists who were quick to apply isozyme techniques to such widely ranging disciplines as clinical chemistry, comparative enzymology, and genetics.

Ten to twelve years ago only a handful of enzymes were shown to occur as isozymes by means of histochemical staining of the electrophoretically separated forms. Today, the number exceeds fifty, and the list continues to grow dramatically as more and more laboratories begin to recognize the importance of isozyme studies to their field of work. In this respect, one of the most active laboratories has been that of Dr. George J. Brewer at the University of Michigan. Dr. Brewer has made extensive use of the study of isozymes relative to his main interests of cellular enzymology and metabolism and biochemical and medical genetics. It is out of this firsthand experience with a broad variety of isozymic techniques, as well as his vigorous curiosity concerning the reasons which could underlie the intriguing diversity of isozyme variation he was observing, that the foundations of this book are based. The results of this effort are extremely satisfactory.

The usefulness of this book is essentially threefold: It admirably organizes and synthesizes the subject matter of this relatively young subdivision of enzymology which heretofore was scattered in various scientific journals or books dealing with limited aspects of isozymes; it brings together in one volume an impressive number of techniques for the analysis and interpretation of isozymes; and, finally, it provocatively discusses the applications and possible significance of isozymes as they relate to the problems of clinical medicine, genetic selection, and organic evolution.

RICHARD E. TASHIAN

Copenhagen, Denmark
January, 1970

Preface

Repeated requests from workers in numerous laboratories for methodological details of various isozyme techniques being performed in our laboratory have stimulated the writing of this book. An obvious need exists to bring together in one place the methods required to perform electrophoretic analysis of a large number of enzymes from a wide variety of organisms. It is hoped that this work will fill that need.

It is intended for several types of workers and laboratories. Specifically, clinical pathologists, clinical biochemists, and physicians should find it valuable for strengthening and enlarging the scope of isozyme studies in clinical disease. A multienzyme approach to clinical problems is emphasized. Zoologists, botanists, geneticists, biochemists, physiologists, microbiologists, systematists—scientists from almost every biological discipline—are increasingly turning to isozyme approaches for the study of numerous types of problems. This book should orient new workers to possible applications of isozyme techniques to their field and provide a variety of specific isozyme techniques for both new and experienced workers. Because of the rapid expansion of isozyme studies, many workers who wish to use these techniques are not well-grounded in biochemical methods. Therefore, early chapters present basic methods in sufficient detail so that even the most inexperienced worker should be capable of initiating isozyme studies. Last, but not least, this book is intended to assist that unsung hero, the laboratory technician. This book should more often than not end up in the laboratory, by the side of the person actually carrying out methods.

Besides chapters covering electrophoretic methods, a chapter has been written in collaboration with Charles F. Sing (Chapter 7) devoted to methods for data acquisition from electrophoretic patterns and data analysis. Two chapters are devoted to a general discussion of isozymology, the strengths, history, principles, terminology (Chapter 1), and present and future applications (Chapter 6).

I am very grateful to my secretary, Mrs. Lynne Bowbeer, for her very considerable effort in preparing the manuscript for this work. I am also much indebted to Mr. David Bowbeer, Assistant in Research, who assisted in the development of many of the methods and proofread parts of the manuscript. I express my appreciation also to others who have worked in my laboratory and who have assisted in the accumulation of improved electrophoretic methods, including Mr. Gary Eaton, Mr.

Conrad Knutsen, Miss Lucia Feitler, Dr. John Eaton, Miss Carol Coleman, and Mr. Dinu Patel. Numerous conversations with Dr. Richard Tashian and Dr. Charles Shaw have provided continuing stimulation for work with isozymes.

I thank the authors whose names appear in the legends and the following editors and publishers who gave permission to reproduce illustrations: The Editor, *Proceedings of the National Academy of Sciences*, Figures 11, 15, 16, and 17; The Editor, *Biochemical Genetics*, and Plenum Press, Figure 12; The Editor, *Brookhaven Symposia in Biology*, Figure 14.

I wish to thank the staff of Academic Press for their assistance and patience. I also wish to acknowledge the continuing and reliable support for our laboratory of the National Institutes of Health, without which the development of the techniques presented would not have been possible.

Chapter 1
Introduction

A. Strengths of the Isozyme Approach

The use of electrophoretic techniques to separate multiple molecular forms of enzymes, with subsequent histochemical staining of isozymes, has virtually exploded onto the biological science scene in the last decade. This approach owes its spreading popularity to several inherent characteristics and strengths.

1. Direct Visualization of Gene Products

One strength of this approach is its capacity to visualize enzymes directly. This allows for the detection of multiple gene products catalyzing the same reaction. It also allows for the detection of genetic variation irrespective of the effect of mutation on quantitative activity.

2. Specificity

The commercial availability of high quality substrates lends many of the isozyme methods considerable specificity. Surprisingly, a large number of such specific reactions are catalyzed by multiple forms of enzymes. An understanding of the mechanisms by which this great isozymic diversity is brought about and of the functions its subserves is important to the study of modern molecular biology.

3. Tissue and Ontogenetic Variation

The considerable differences in isozymes among tissues of most organisms, and during the life cycle, are also important strengths of isozyme studies. These differences can be used clinically to detect organ damage or to study tissue differentiation and ontogeny.

4. Simplicity

A great advantage of the isozyme approach is that studies can be carried out on crude extracts. Purification of the enzymes is not required, greatly simplifying matters, and placing such studies within reach of laboratories of all types.

In this connection, the beginning worker, or the experienced worker in other fields planning to initiate work with isozymes, should not be intimidated by the apparent complexity of the field. Many factors are involved in electrophoresis, but for the most part it is not necessary to have more than a rudimentary knowledge of various principles in order to use the approach successfully. The techniques may be considered quite "robust." A variety of combinations of electrophoretic media, buffers, pH, ionic strength, etc., will usually work satisfactorily. As a matter of fact, most of the buffer systems and other details of the specific methods have been developed quite empirically, by trial and error. In this connection, published methods, and the methods presented in this book, should not be considered necessarily ideal. Further experimentation with and modification of a method can often result in significant improvement.

5. Isozymes as Labels

Isozymes and genetic variation of isozymes serve as "labels" or "markers" which can be used in the study of cultured cells, in linkage studies, and in population studies. Because the techniques require only a small amount of unpurified material, they are easily used in mass screening.

6. Isozymes as Biochemical Tools

The zymogram method lends itself well to the analysis of subunit structure and the demonstration of multiple substrate specificity of enzymes. The method may also reveal previously unsuspected structural or catalytic relationships between groups of enzymes.

B. History

This section is designed primarily to put isozyme techniques in proper perspective with past developments in the field and, coupled with Chapter 6 (on present applications and the future of isozymology), should give the interested reader some orientation in the field's prior accomplishments, and some feeling for the directions in which the field may be moving.

To place the development of isozyme techniques in proper historical perspective, early work leading to the separation of nonenzyme proteins by electrical current must be considered. Tiselius (1937), who may be considered the father of electrophoresis, developed the "moving boundary" method to analyze serum proteins in solution. He was able to distinguish 5 fractions in aqueous solution by "schlieren" optics which identify refractive index gradients. The term "moving boundary" refers to the displacement of the leading edge of the particular protein fraction. By modern standards, this is not a good method for complete separation of fractions because it is technically difficult and requires relatively large samples. Nevertheless it established the principle by which later methods developed.

Subsequent developments involved utilization of "zone" electrophoresis, in which each protein component was separated into a distinct zone in a stabilized media, rather than in solution. Although it is possible to stabilize solutions to a certain extent by the use of an electrophoretically immobile solute such as sucrose, it is far more common to employ medium with a solid matrix such as filter paper, cellulose acetate, agar gel, acrylamide gel, and starch gel. The interstices of these matrices contain the buffer which transmits the electric current, and through which the proteins migrate. Detection of the protein zones by refraction in such media is not possible due to light scattering. Modern methods of detecting enzymes after electrophoresis almost universally depend upon histochemical techniques in which the reaction product of the enzyme leads, in some manner, to a colored area or stain. This approach was first used by Hunter and Markert (1957).

Because of the importance of developing appropriate media for zone electrophoresis, the important historical features in isozymology consist, in large part, of a description of the introduction of different media.

Since the use of filter paper as a supporting media for zone electrophoresis of serum proteins was well established by the mid-1950's, it was only natural that this was the first media used for studies of isozymes. Early work with paper electrophoresis of enzymes included the study of alkaline phosphatase by Baker and Pellegrino (1954), and lactic dehydrogenase (LDH) by Wieland and Pfleiderer (1957), and Sayre and Hill (1957). Heterogeneity of LDH was actually first recognized by Neilands (1952) who found LDH activity in two electrophoretically distinct protein fractions separated earlier by Meister

(1950). It was not long after studies of LDH isozymes were initiated that the isozyme composition of the serum was found to be of value in clinical medicine (Vesell and Bearn, 1958; Hess, 1958). Hess used paper, while Vesell and Bearn used starch block as media.

Gel electrophoresis first involved agar gels made up with a buffered solution. Separation of proteins by agar gel electrophoresis was first accomplished by Gordon et al. (1949), and was first applied to the separation of isozymes by Wieme (1959).

The use of starch grains in so-called starch block electrophoresis was first introduced by Kunkel and Slater (1952), and the first isozyme studies were accomplished by Vesell and Bearn (1957) on lactate and malate dehydrogenases. Starch block electrophoresis has been found useful for "preparative" electrophoresis, that is, separation of a sufficient quantity of protein for further analysis. This type of electrophoresis contrasts with the usual "analytical" electrophoresis in which the goal is simple identification of banding patterns.

In 1955 Smithies introduced the technique of starch gel electrophoresis. This technique has seen wider use in the study of isozymes than any other method. Powdered starch is hydrolyzed by heating in an aqueous solution. As the solution cools it forms a gel. This type of gel (also acrylamide gel) contributes an additional factor to the separation of proteins that is absent in other media. The pores in the gel matrix are of the same order of magnitude in size as the protein molecules, resulting in a "molecular sieving" effect which further contributes to protein separation. This property leads to greater sensitivity in the resolution of protein bands.

Starch gel was first used in enzyme studies by Hunter and Markert (1957) who suggested the term "zymogram" to describe the starch gel strips stained by histochemical methods. Markert and Moller (1959) later suggested the word "isozyme" to describe the enzyme bands which appeared after staining.

Cellulose acetate strips have been found to be suitable for many types of electrophoresis. Kohn (1957) first reported the use of cellulose acetate as a medium for electrophoresis of proteins. It was first used for the study of lactate dehydrogenase isozymes by Wieland et al. (1959).

Acrylamide gel electrophoresis was introduced in 1959 by Raymond and Weintraub (1959), and Ornstein and Davis (1959). In spite of its recent introduction, this medium has already seen considerable use because it has excellent resolving power. Most gels, including acrylamide, are poured in slabs. However, one variation of the acrylamide method involves pouring the gel in cylindrical glass tubes; this has led to the term "disc" electrophoresis.

An important landmark in the short history of isozymology was the demonstration of inherited electrophoretic variation. Early reports were those of Allen (1960) on esterases of *Tetrahymena*, Schwartz (1960) on esterases of maize, Tashian (1961) on esterases of human erythrocytes, Boyer (1961) on human alkaline phosphatase, and Porter *et al.* (1961) on human glucose-6-phosphate dehydrogenase (G-6-PD).

Tashian's laboratory has also made a number of other important contributions. They demonstrated that a subgroup of red cell isozymes, initially detected by a strain for esterase activity, were actually carbonic anhydrases (Tashian and Shaw, 1962). This illustrated the principle that the true physiological roles of isozymes detected by nonspecific methods can be identified, given time and perseverance. Using the esterase system. Tashian (1965, 1969) has also demonstrated that the component enzymes of an isozymic group, while differing in molecular weight and kinetic characteristics, can share subunits. This was shown by differences in migration of all 3 groups of A esterase of human erythrocytes as a result of a single gene mutation. The first demonstration of an amino acid substitution in an electrophoretically detected mutant enzyme was also reported by Tashian *et al.* (1966), once again involving carbonic anhydrase of human erythrocytes. Tashian's findings indicate that the same principles of gene action that applied to inherited protein variants such as hemoglobin can be applied to enzymes of diploid organisms.

The first demonstration of a genetic polymorphism involving electrophoretically different enzymes of man was that of Porter *et al.* (1961) involving erythrocytic G-6-PD in Negro populations. Numerous other polymorphisms of electrophoretically different enzymes have been reported in a great variety of organisms. The significance of the high frequency of electrophoretically detected polymorphism was not at first appreciated. Shaw (1964, 1965) seems to have been the first to take note of this high frequency. It has been pointed out by a number of authors that the isozyme technique allows a sampling of genetic loci (by sampling enzymes) in a manner which at least approaches randomness. By determining the proportion of randomly selected enzymes which show electrophoretic variation in a population, an investigator can make a rough estimate of the proportion of the genetic loci of the population which shows variation. Before electrophoretic techniques, studies of available genetic systems could not allow a realistic estimate because of the highly biased way in which the existence of the various gene products came to the attention of the investigator. For example, considerable genetic variation has been detected in the blood groups of man, but it is not possible to know what proportion of blood group loci show variation because the total number of loci involved is unknown; those blood group systems with variation come to attention, but those without variation go undetected.

Lewontin and Hubby (1966), and Harris (1966) have utilized the electrophoretic approach as outlined above. Lewontin and Hubby (1966) report that 39% of the loci of natural populations of *Drosophila pseudoobscura* are polymorphic. Harris (1966) finds that 30% of human erythrocyte enzymes show polymorphic variation. These results have caused widespread reverberations in the area of population genetics. Previous theory had suggested that a rather low upper limit may exist for the number of polymorphic loci a population can maintain. Following the publication of the isozyme data, however, a spate of papers appeared (Sved *et al.*, 1967; King, 1967; Milkman, 1967) which introduced models, reintroduced the arguments of Sanghvi (1963) and Li (1963), and could account for the maintenance of the great amount of polymorphism observed by the isozyme techniques.

Another important application of isozyme techniques was the use of the glucose-6-phosphate dehydrogenase (G-6-PD) system in the study of X-chromosomal inactivation. Lyon (1961), based on her work with X-linked coat color mutants in mice, postulated that only one X-chromosome was active in each female mammalian cell. G-6-PD is controlled by an X-linked gene in man, and an electrophoretic polymorphism exists in the Negro. Davidson *et al.* (1963) cloned fibroblasts from women who were heterozygous for the electrophoretic types and showed that colonies derived from single cells expressed only one or the other, never both, electrophoretic phenotypes. This demonstrated that in each female cell only one G-6-PD allele was active.

C. Principles

It is beyond the scope of this book to give a full theoretical treatment to the principles underlying electrophoresis. This has been done admirably elsewhere, for example in the recent book edited by Bier (1967). Nonetheless, it is of some importance for individuals working with isozymes to have at least a general concept of the principles involved in the separation of proteins by electric current.

In general, the principle of electrophoretic separation of proteins depends upon passing an electric current through an electrophoretic media, and upon the possession of varying electrical charge by the proteins to be separated. In order to transmit current, the electrophoretic media must contain an ionized solution or buffer.

The type of buffer to be used will vary with the system. In general, the use of specific buffers for specific isozyme systems has been built up empirically by trial and error. The ionic strength of the buffer solution has effects in electrophoresis that the investigator should be aware of in perfecting a new system. Buffers of low ionic strength permit fast migration and produce relatively low heat development, while buffers of higher ionic strength give, in general, sharper separation of zones. The optimal pH for the buffer will vary according to the charge characteristics of the proteins to be studied. The charge on a protein molecule will depend upon the proportion of its carboxyl and amino groups which are charged. Carboxyl groups develop a negative charge and amino groups a positive charge. At its isoelectric point a protein is neutral, that is, it has the same number of charged carboxyl and amino groups. As the pH is decreased, the amino groups are progressively ionized and the protein assumes a positive charge. As the pH is increased, the carboxyl groups are progressively ionized, giving the protein a more negative charge. The net electrical charge on a protein at a given pH will depend upon the number of exposed amino and carboxyl groups which are ionized. The greater the charge on a protein, the faster it will move in an electric field in the direction of the electrode having the opposite charge.

Besides the effect of the net electric charge of the protein, separation can also be affected by molecular sieving in the electrophoretic medium. That is, due to the fact that certain electrophoretic media, notably starch and acrylamide, have a pore size approaching that of protein molecules, the size and shape of the protein molecule will affect its rate of migration. Another factor which may affect migration is the combination of ions in the electrophoretic medium with uncharged groups of the protein to form charged complexes.

In general, the rate of migration of a protein is dependent upon the strength of the electric current. However, with certain electrophoretic media, notably paper, a phenomenon known as electroosmosis may also influence migration. When paper is exposed to an electric field, the carboxyl groups of the paper become ionized, and this induces a positive charge on solvent molecules which then migrate towards the cathode. This electroosmotic flow of the solvent will tend to carry the proteins toward the cathode. The extent of this movement can be evaluated with a nonionized dye marker placed on the paper. Alkaline buffers tend to increase ionization of the carboxyl groups on paper, and thus accentuate electroosmosis. Electroosmosis is minimized by increasing buffer ionic strength.

D. Terminology and Definitions

The comments in this section are offered primarily as suggestions and as evidence of the author's concern that some guidelines to isozyme nomenclature begin to be developed.*

1. Isozymes versus Isoenzymes

Some workers prefer the term isozyme, others isoenzyme. Usage dictates that the two are interchangeable. The term isozyme will be used in this book.

2. Definition of Isozymes

Most workers would agree that isozymes have at least the following minimal characteristics: they are multiple molecular forms of enzymes derived from the same organism and have at least one substrate in common. Beyond this, a great variety of opinions have been expressed on further restrictions to place on the definition of "true" isozymes. There are those who feel that the term should be restricted to those multiple molecular forms of enzymes which are derived from the same organ or tissue, have similar genetic origins, and have very similar, not just overlapping, catalytic activities. Then, of course, all possible combinations of these constraints have been suggested. The idea behind a more restrictive definition is to try to limit the definition to closely related enzymes, with similar genetic origins, and closely related biochemical specificities such as those manifested by the LDH isozymes. For example, some have felt that multiple molecular forms of certain enzymes from specific organs or tissues which show broad substrate specificities, such as phosphatases, peroxidases, and esterases, do not qualify for inclusion within the definition of isozymes because certain enzymes of

---

* It should be noted that the problems of nomenclature of isozymes are beginning to attract attention. The Ad Hoc Committee on Nomenclature of Isozymes of the Isozyme Conference, of which the author is a member, has recently communicated a set of recommendations to the Office of Biochemical Nomenclature of the Division of Chemistry and Chemical Technology, National Research Council. Recommendations on isozyme nomenclature may be forthcoming in the near future from this office.

these groups may have had entirely different origins. Thus, a definition requiring merely that isozymes come from the same organ and share catalytic activities does not satisfy these individuals. Furthermore, it is now known that different isozymes may exist within different intracellular compartments, such as those associated with mitochondria versus the soluble cytoplasm. If one is to define isozymes on the basis of tissue specificity, it would appear that the role of intracellular differentiation should also be considered.

Some workers have suggested that the definition should require isozymes to have the same coenzyme. Aside from the fact that many enzymes do not have coenzymes, this restriction can also lead to considerable confusion since it has been shown that enzymes sharing a catalytic function and using the same coenzyme, but with otherwise widely differing properties, exist within the same cell. Furthermore, many enzymes are capable of using more than one coenzyme.

It is our opinion that definitions of various restrictive types are unnecessary. With all due respect to Markert and Moller who coined the word isozyme in 1959, it is difficult to define a word contrary to well-established usage. It seems obvious that the principle usage of the word isozyme is an operational one. That is, isozymes, by usage, are multiple molecular forms of enzymes seen after separation procedures such as electrophoresis. Multiple forms separated by chromatography and other such methods have also been referred to as isozymes. The multiple forms are almost invariably derived from the same organism; this seems a useful criteria for isozymes which fits with usage. Even this restriction may not be pertinent, however, if one considers electrophoretic studies of mixed cell cultures which have hybridized. Isozymes are, of course, almost always revealed by some type of procedure (such as histochemical) which indicates a shared catalytic activity. Our definition of isozymes then, would be multiple molecular forms of enzymes derived from the same organism (or tissue culture) sharing a catalytic activity.

In the great majority of cases, when an investigator sees a number of enzyme bands after an electrophoretic run with a new system or new population of organisms, he will not be in position to decide if the enzymes are related genetically, structurally, or biochemically. Such dissection of the isozyme pattern often takes many years. It is not reasonable to hold the system in limbo until such investigations have been completed, nor is it practical to call a group of enzymes isozymes for several years and then stop calling them isozymes because they are found to be different in one or more arbitrary characteristics. As a matter of fact, it would not be at all surprising if most of the so-called nonspecific isozymes, those revealed by shared activity on a synthetic substrate, were evolutionarily related. Thus, in our opinion it is quite appropriate for the investigator to call a newly discovered pattern an isozyme pattern (or zymogram) and to call the bands he sees isozymes. Differentiation of the types of isozymes should come from the development of new terms, not redefinition of old ones.

3. Types of Isozymes

a. Genetically Determined Electrophoretic Variation

If a mutation results in a substitution of an amino acid which changes the net charge on a protein, electrophoretic migration will be affected. Since many amino acids are neutral, and because a mutation can result in a substitution of a like-charged amino acid, it is apparent that not all mutations will be electrophoretically detectable. According to calculations (MacCluer, cited in Shaw, 1965), about one-third of all possible amino acid substitutions should be detectable by electrophoresis. However, this is a theoretic figure and it should be borne in mind that if the tertiary (folding) structure of the protein is affected by a mutation, it may migrate at a different rate, even if the substituted amino acid should not have led to a charge difference.

The question has occasionally been raised whether enzymes that show different electrophoretic migration because of allelic variation should be called isozymes. It seems to us that if such allelic variation results in multiple molecular forms of enzymes within an organism, they are isozymes. For example, LDH commonly shows a 5-band isozyme pattern in mammalian tissues. The pattern is common to all organisms in the population. These isozymes arise from randomly combining tetramers of two types of polypeptide subunits. If an electrophoretically detectable mutant occurs in one of the subunits, a 15-band pattern is generated (Shaw and Barto, 1963). It seems entirely appropriate to designate all 15 bands, including the 10 new ones, as isozymes. In our laboratory we consider bands which vary from one individual to another in a population as a subdivision of isozymes and call them "segregating isozymes" or "segrazymes."

b. Nonsegregating Isozymes

In discussions of isozyme patterns in which segregating (segrazymes) and nonsegregating enzyme bands are being considered, it is convenient to have a term such as "nonsegregating isozymes" to designate those bands common to all members of the population. We have also used the term "constazymes" for this purpose. The usual 5-band LDH pattern in a given species of mammal, then, would consist of nonsegregating isozymes or constazymes.

c. Homopolymers

A protein consisting of more than one identical subunit is a homopolymer. Variation in the size of a molecule resulting from variation in the number of identical subunits can produce isozymes, particularly in media which have a sieving effect, such as starch gel and acrylamide gel.

d. Heteropolymers

A protein consisting of two or more types of subunits is a heteropolymer. Isozymes can result from varying combinations of such subunits, such as the LDH example cited abve. Combinations of subunits may be random, as with LDH, nonrandom, as with hemoglobin. Heteropolymers formed by *in vitro* manipulation are sometimes called "hybrid" molecules, a term which should probably be abandoned in favor of heteropolymer, unless some type of actual biological hybridization has taken place.

e. Conformational Isozymes

Multiple molecular forms of proteins may occur in which the primary structure (amino acid sequence) is identical in all of the forms. This has been attributed to conformational, or folding, differences in the tertiary structure. If a protein folds differently, a different proportion of charged amino or carboxyl groups may be exposed, resulting in a difference in net charge. These isozymes have been called "conformational isozymes" or "conformers."

The cause or causes of conformational isozymes are unknown. Many workers view them as artifacts arising from differential binding of small molecules during *in vitro* preparative procedures. Indeed, it has been amply demonstrated that conformational isozymes can be generated *in vitro*. However, it is also possible that such isozymes are biologically relevant. If they do occur *in vivo*, it may be one of the many possible mechanisms for intracellular organization of enzymes (see Chapter 7).

f. Isokinetic Isozymes

The term "isokinetic" has been applied to isozymes which show approximately equal quantitative activity. It should be kept in mind when using this term that the activity measurements are always carried out *in vitro* with high concentrations of substrate and coenzymes. Such measurements may not accurately reflect the *in vivo* activities of isozymes.

4. Numbering of Isozyme Bands

It is convenient to have a system of numbering (or lettering) isozyme bands for easy reference to specific bands as, for example, in a publication. By using numbers (or letters) we eliminate the problems that develop when 2 bands are initially called "fast" and "slow," then additional variants are called "faster" or "slower," and from there on chaos develops as intermediate bands, or very slow or very fast migrating bands, are observed.

It is generally agreed that the fastest migrating anodal band should be numbered "1" (or "a") and numbering should be continued in a cathodal direction passing through the origin and ending with the cathodal end of the gel. It is also becoming accepted that in figures and publications the anode of the gel should be either at the top or at the right-hand side of the figure.

It is our view that in initial studies, the numbers assigned to bands should be tentative and primarily applicable only to that particular publication. In other words, the numbers placed on isozyme bands in photographs or diagrams in manuscripts should not have much more permanent relevance than the figure number of that particular photograph. The numbers should be used only for description and easy reference. The reasons for adopting a flexible numbering system include changes in band positions with change in pH and buffers, and the invariable occurrence that subsequent variants include faster and slower isozymes.

If, for example, isozymes are numbered 1–7 initially (Figure 1, slot 1) and thereafter a faster band is observed, and if a flexible numbering system is in use, in the subsequent publications showing the new bands, all the bands may be renumbered (Figure 1, slot 2). Alternately, a regional labeling method may be used with retention of the original band numbers (Figure 1, slot 3). A regional approach consists of dividing the pattern into regions, with letters designating the regions, and numbers designating the bands within a region. Once again, if future developments require, it would be possible to revise the regional nomenclature.

Permanence may come to isozyme nomenclatural designations in several ways. If a numbering system eventually covers the needs of the system, such as with LDH, it may become permanent through usage. Once the genetics of a given part of the system have been established the phenotypic designations should be permanent, at least until the amino acid sequence has been established, in which case specific structural designations should then be used. The phenotypic designations can be letters or numbers initially, but after the first description of a system it is probably better for subsequent names to be trivial, as with the hemoglobin system. The major purpose of this system is to avoid duplication in naming variants.

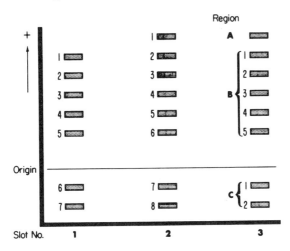

Figure 1. Diagram illustrating the possible evolution of an isozyme nomenclatural system. Slot 1 shows the numbers assigned to the bands observed in the original study. Slot 2 shows a possible renumbering later when a faster band is observed. Slot 3 shows a possible relabeling system based on a regional approach.

As pointed out in the report of the WHO Scientific Group on Standardization of Procedures for the Study of Glucose-6-Phosphate Dehydrogenase (Standardization of Procedures, 1967), after a system begins to evolve it is very difficult to distinguish genetic variants. Electrophoretic migration, even with assay of quantitative activity, is inadequate to distinguish variants. At this point it becomes increasingly important to characterize variant enzymes by additional studies. In the case of glucose-6-phosphate dehydrogenase (G-6-PD) the report suggested that 5 characteristics be evaluated before adjudging an alledgedly new variant unique.

With complicated enzyme systems, meetings of the type mentioned above for the G-6-PD system should eventually be held by the investigators most involved and agreement should be reached on nomenclatural problems within the system.

REFERENCES

Allen, S. (1960). *Genetics* **45**, 1051.
Baker, R. W., and Pellegrino, C. (1954). *Scand. J. Clin. Invest.* **6**, 94.
Bier, M., ed. (1967). "Electrophoresis, Theory, Methods and Applications," Volume 2. Academic Press, New York.
Boyer, S. H. (1961). *Science* **134**, 1002.
Davidson, R. G., Nitowsky, H. M., and Childs, B. (1963). *Proc. Natl. Acad. Sci. U. S.* **50**, 481.
Gordon, A. H., Keil, B., and Sebesta, K. (1949). *Nature* **164**, 498.
Harris, H. (1966). *Proc. Roy. Soc. (London)* **B164**, 298.
Hess, B. (1958). *Ann. N. Y. Acad. Sci.* **75**, 292.
Hunter, R. L., and Markert, C. L. (1957). *Science* **125**, 1294.
King, J. (1967). *Genetics* **55**, 483.
Kohn, J. (1957). *Biochem. J.* **66**, 9P.
Kunkel, H. G., and Slater, R. J. (1952). *Proc. Soc. Exptl. Biol. Med.* **80**, 452.
Lewontin, R. C., and Hubby, J. L. (1966). *Genetics* **54**, 595.
Li, C. (1963). *Am. J. Human Genet.* **15**, 316.
Lyon, M. (1961). *Nature* **190**, 372.
Markert, C., and Moller, F. (1959). *Proc. Natl. Acad. Sci. U. S.* **45**, 753.
Meister, A. (1950). *J. Biol. Chem.* **184**, 117.
Milkman, R. (1967). *Genetics* **55**, 493.
Neilands, J. B. (1952). *J. Biol. Chem.* **199**, 373.
Ornstein, L., and Davis, B. J. (1959). "Disc Electrophoresis." Distillation Products Industries (Division of Eastman Kodak Co.).
Porter, I. H., Boyer, S. H., Schulze, J., and McKusick, V. A. (1961). *Proc. 2nd Intern. Conf. Human Genet., Rome.* (Abstr.)
Raymond, S., and Weintraub, L. (1959). *Science* **130**, 711.
Sanghvi, L. (1963). *Am. J. Human Genet.* **15**, 298.
Sayre, F. W., and Hill, B. R. (1957). *Proc. Soc. Exptl. Biol. Med.* **96**, 695.
Schwartz, D. (1960). *Proc. Natl. Acad. Sci. U. S.* **46**, 1210.
Shaw, C. R. (1964). *Brookhaven Symp. Biol.* **17**, 117.
Shaw, C. R. (1965). *Science* **149**, 936.
Shaw, C. R., and Barto, E. (1963). *Proc. Natl. Acad. Sci. U. S.* **50**, 211.
Smithies, O. (1955). *Biochem. J.* **61**, 629.
Standardization of Procedures for the Study of Glucose-6-Phosphate Dehydrogenase (1967). *World Health Organ. Tech. Rept. Ser.* **366**.
Sved, J., Reed, T., and Bodmer, W. (1967). *Genetics* **55**, 469.
Tashian, R. E. (1961). *Proc. Soc. Exptl. Biol. Med.* **108**, 364.
Tashian, R. E. (1965). *Am. J. Hum. Genet.* **17**, 257.
Tashian, R. E. (1969). In "Biochemical Methods in Red Cell Genetics" (J. Yunis, ed.). Academic Press, New York, pp. 307–336.
Tashian, R. E., and Shaw, M. W. (1962). *Am. J. Hum. Genet.* **14**, 295.

Tashian, R. E., Riggs, S. K., and Yu, Y. S. L. (1966). *Arch. Biochem. Biophys.* **117**, 320.
Tiselius, A. (1937). *Trans. Faraday Soc.* **33**, 524.
Vesell, E. S., and Bearn, A. G. (1957). *Proc. Soc. Exptl. Biol. Med.* **94**, 96.
Vesell, E. S., and Bearn, A. G. (1958). *J. Clin. Invest.* **37**, 672.
Wieland, T., and Pfleiderer, G. (1957). *Biochem. Z.* **329**, 112.
Wieland, T., Pfleiderer, G., and Ortanderl, F. (1959). *Biochem. Z.* **331**, 103.
Wieme, R. J. (1959). "Studies on Agar-Gel Electrophoresis." Arscia, Brussels.

Chapter 2
Starch Gel Electrophoresis

A. Introduction

Although general descriptions are given for the use of a number of electrophoretic media in this book (see Chapter 3) most weight has been placed on starch gel electrophoresis, to which this entire chapter is devoted. The rather heavy emphasis on starch gel compared to other media is the result of two factors. First, it is the method which the author uses primarily in his own laboratory. Second, the author believes that, as of the moment, starch gel is the best general media for isozyme studies. We have found that it has a greater resolving power for isozymes than other media, including acrylamide, and, in addition, serves effectively for large scale screening. Starch gel electrophoresis has some disadvantages. Unless the gel is specially processed it is not clear, making direct densitometric studies somewhat more difficult than with transparent media. Starch gels are somewhat friable and not easy to handle. In general, starch gel electrophoresis requires a longer electrophoretic run than some of the other methods. However, it is our opinion that some of the limitations of other methods are more disadvantageous (as discussed in Chapter 3). In many types of laboratories, however, one or more of the other methods described in Chapter 3 may be quite profitably combined with starch gel electrophoresis, and this may result in some conservation of time or materials.

Irrespective of the type of media to be used in a given laboratory, a good understanding of the use of starch gel and the related equipment will make it fairly easy to use the other media. In general, the specific methods described in Chapter 5, while developed primarily with the use of starch gel, should lend themselves fairly readily to other media, particularly gel media such as acrylamide and agar.

In order to carry out starch gel electrophoresis a starch gel must

B. Starch Gel Trays and Related Parts / 17

first be prepared in the laboratory from powdered starch. The process consists of making a solution of starch, cooking it, and then pouring it into a mold, called a starch gel tray. As the starch solution cools, it solidifies into a gel which is used for electrophoresis. The presentation in this chapter is oriented primarily toward vertical electrophoresis, which gives somewhat better resolution than horizontal electrophoresis. However, in most cases the descriptions are pertinent to either type.

B. Starch Gel Trays and Related Parts

1. Construction

The starch gel tray has several component parts as shown in Figure 2. Trays for starch gel electrophoresis can be of a variety of sizes and

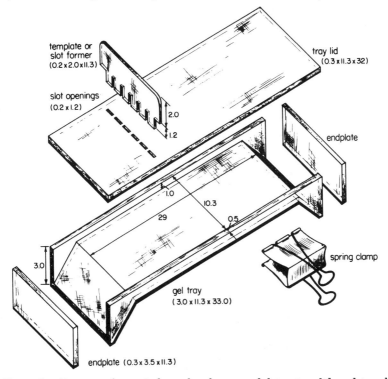

Figure 2. Diagram of a typical starch gel tray, endplates, tray lid, and template. Numbers indicate dimensions in cm. See also Table I for dimensions of starch gel equipment. One spring clamp is shown. Typically 3 such clamps would be used on each side of the tray.

types. The types employed in our laboratory are made of plexiglass, constructed according to our specifications by local companies or persons. (Equipment for vertical gel electrophoresis is also commercially available from Buchler Instruments, Inc., Fort Lee, New Jersey, E-C Apparatus Corp., Philadelphia, Pennsylvania, and Otto Hiller, P.O. Box 1294, Madison, Wisconsin.) Once a suitable tray has been constructed it is wise to send the original along as a model when ordering additional trays of the same type, so that all trays, end plates, tray lids, and templates will be interchangeable. Alternately, the company may keep the specifications for each tray type on file. Plexiglass is a suitable material for construction of the trays because it is a nonconductor of electricity, inexpensive, and easily worked with; it is transparent, thus permitting visualization of the gel inside, and the starch gel does not stick to its surface.

Figure 2 is a diagram which demonstrates a typical starch gel electrophoretic tray, end plates, tray lid, and template. Figure 3 is an actual photograph of a gel after it has been poured, prior to removal of template, lid, and endplates. The names of the various pieces of the gel tray are indicated in Figure 2, and this figure should be consulted as necessary to understand the following sections.

Various modifications of starch gel electrophoresis have been described, including thin-layer starch gel electrophoresis on glass slides (Daams, 1963; Ramsey, 1963) and micro starch gel procedures (Boivin et al., 1959; Mouray et al., 1961; Marsh et al., 1964; Koch et al., 1964).

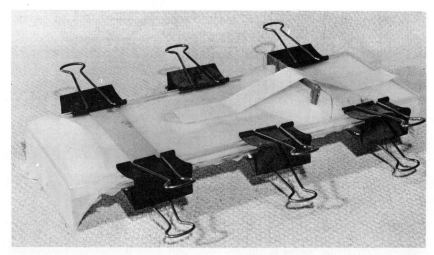

Figure 3. Photograph of a starch gel in a starch gel tray prior to removal of template, endplates, clamps, and lid.

2. Dimensions

A reasonable length for starch gel trays has been determined empirically on the basis of experience. The most important feature in this regard is that the trays be sufficiently long so that after a period of electrophoresis sufficient to separate slowly migrating isozymes, faster migrating isozymes have not migrated completely or partially off of the starch gel. On the other hand, the gel should be sufficiently short so that a good flow of electrical current occurs and the gels are reasonably easy to handle.

The depth of the starch gel tray determines the thickness of the resulting gel. The thicker it is, the more easily it can be cut into a number of horizontal slices after electrophoresis, and the larger the number of enzymes which can be analyzed after a single electrophoretic run. However, the gel must not be so thick that it becomes too warm in its interior during electrophoresis. The electrical current causes warming of the gel which cools at its surface and, if it is too thick, may become very warm at its center. Undue heating may cause loss of enzyme activity and distortion of isozyme patterns. In our laboratory we employ gel thicknesses of either 1.0 or 0.6 cm. It is always possible to obtain 2, usually possible to obtain 3, and sometimes possible to obtain 4 slices from a gel of 1 cm thickness. If it is desired not to slice the gel into more than 2 parts, a gel tray of 0.6 cm depth may be used. With starch gel, it is always necessary to slice the gel at least once, since the outer surfaces do not stain properly. The ends of the starch gel tray are shaped so as to provide a thicker, usually triangular, section of gel, increasing the surface area at the ends of the gel where the current is transmitted (Figure 2).

The width of the starch gel tray will vary according to the needs of the investigator. In general, the wider the tray, the larger the number of samples that can be studied on the same gel. It is an advantage in comparative studies to study samples on the same gel because this eliminates gel-to-gel variation. With a wide starch gel, and a large number of slots, many contrasts can be made on the same gel. On the other hand, wider gels are more difficult to handle than smaller gels.

3. Tray Lids

We must consider the number of slots and the length of the sample slots (the long axis of the slot is regarded as the slot length) in the tray lid. In general, the shorter the slots, the larger the number of slots which can be accommodated in a gel tray of a given width. Thus, short

Table I
NUMBER OF SLOTS AND DIMENSIONS OF REPRESENTATIVE TYPES OF STARCH GEL TRAYS AND TRAY LIDS

No. of slots	Internal width of gel tray (in centimeters)	Slot length (in centimeters)	Distance between slots (in centimeters)
6	10.3	1.2	0.3
8	10.3	0.8	0.3
12	10.3	0.7	0.1
30	25.5	0.6	0.2

slots and wide trays are advantageous in large scale screening programs. On the other hand, longer slots give better banding characteristics and lend themselves well to detailed investigation of isozymes. Table I lists the dimensions and slot numbers of the 4 types of gel trays and tray lids used in our laboratory. These tray dimensions span a variety of uses and can serve as a general guide. Note that 2 basic gel trays are used (internal width 10.3 and 25.5 cm) with modifications made possible by the use of varying types of tray lids and templates. The length of the trays are all identical, 33 cm, as shown in Figure 2.

4. Templates

Templates (also called slot formers or combs) are designed in such a way as to penetrate through the sample slot on the gel tray lid and into the starch solution while it is still liquid form (Figures 2 and 3). After the starch solution solidifies into a gel, removal of the template will leave holes in the gel in which the samples to be studied can be placed. The construction of the template is rather critical in several respects. Its teeth must be spaced to fit into the gel lid, and they must be long enough to come within about 1 mm of the gel tray bottom. If the teeth touch the bottom of the gel tray, the samples may run out through the bottom of the gel slots and spread along the interface between gel tray and gel. If the teeth of the template are too short, the surface area of the sample will not be as great, and migration patterns will not be even throughout the thickness of the gel. In addition, the thickness of the template is of some importance because it determines the volume of solution which can be added to the sample slot, and the width of the starting point for the proteins. Obviously, the thicker the teeth of the template, the larger the volume of the sample slot in the gel. The volume of sample in a slot in the gel can be used as a variable to make banding patterns either darker or lighter. Occasionally it is difficult to determine various bands because enzyme activity is excessive. Improvement in resolution in these cases can be obtained by diluting the sample, using a thinner template, or both. On other occasions when banding patterns are too faint, their strength can be increased by using a more concentrated solution, thicker templates, or both. Of course, with thicker templates the starting positions of the protein molecules is broader, which at least slightly decreases resolution. The thickness of the two types of templates used in our laboratory are 0.2 and 0.1 cm. If all tray lids have slots of 0.2 cm width, they may be used interchangeably with either thickness template.

There are other sample insertion methods in addition to the use of templates. After removing the lid (solid, with no holes in this case) cuts may be made in the gel with a razor blade or knife, and filter paper inserts used (as described for agar—see Chapter 3, Section C). Latner and Skillen (1968) describe a starch insertion method in which a small block of gel is cut out by means of two razor blades mounted 2–3 mm apart. The sample under investigation is placed in a suspension of starch grains which is then packed into the slot and covered with molten petroleum jelly.

C. Preparation of the Starch Gel

1. Standard Method

Hydrolyzed potato starch powder from two sources are available and have been utilized in our laboratory. The first is a product of Connaught Medical Research Laboratories, Toronto, Canada, distributed by Fisher Scientific Co., Pittsburgh, Pennsylvania. The amount of Connaught starch to be used for gel electrophoresis varies from lot to lot and is listed on the label of each container. The second type is Electrostarch from the Electrostarch Co., Otto Hiller, P. O. Box 1294, Madison, Wisconsin. Electrostarch has the advantage of forming a gel with increased tensile strength making the gel easier to handle. 10.6 gm of Electrostarch is used per 100 ml of gel buffer. (The amount does not vary from lot to lot.)

To prepare a standard small (6 to 12 slot, 0.6 cm thick) gel, sufficient starch is weighed to make 400 ml of solution and placed in a 1 liter Pyrex side-arm flask. If a larger gel is to be made (such as a 30 slot, 1.0 cm thick gel) sufficient starch is weighed for 1000 ml of solution and placed in a 2 liter side-arm flask. Next, the appropriate amount (400 or 1000 ml) of the particular buffer to be used for a given enzyme procedure (specified in Chapter 5) is added to the starch and swirled until the starch is dissolved. The flask is then placed under a mixer. We have employed a Vari-Speed Mixer purchased from Precision Scientific Co., Chicago, Illinois, but any type of laboratory stirring device which permits vigorous agitation of the contents of the flask can be used.

While the contents of the flask are mixed vigorously, cooking is begun from below with maximum flame from a Bunsen burner. The mixture is cooked for about 4 minutes, but the length of time will vary according to the conditions in the laboratory, such as the amount of heat from the burner, etc. The amount of cooking of the starch solution is quite important. If it is overcooked the gel will be rather fragile; if it is undercooked, it will not provide a uniform matrix for electrophoresis. Unfortunately, there are no really good objective criteria for determining the point at which the starch solution is properly cooked. The art must be learned. Laboratory workers use a variety of techniques for identifying the cooking endpoint. As the starch solutions cooks, it becomes increasingly viscous. Some workers pick up the flask (with an asbestos glove) and can tell by the viscosity of the solution that it is ready. Others observe changes in the interface of the solution and the flask at the top of the solution. Another indication that the gel has been adequately cooked is the formation of small bubbles thoughout the solution. After cooking 5 or 6 gels, a laboratory worker will usually begin to have a fairly good "feel" for the length of the cooking time.

After the gel is cooked, the flask is removed from the flame, a size 8 rubber stopper is placed in the top of the flask, and the side arm is connected to an aspirator. The purpose of this procedure is to degas the solution. If air is allowed to remain in the solution it will form bubbles when the gel solidifies. At the start of the aspiration both large and small bubbles will be seen coming from the starch solution. When the small bubbles stop forming (about 10 seconds) the solution has been adequately degassed. The large bubbles will continue to form because the hot starch solution is actually boiling at the reduced pressure.

The starch solution is poured into the gel tray (with endplates taped in place) so that it overflows on all sides and ends. An excess is poured so that no air pockets will remain when the lid is placed on the tray. It is convenient to place the gel tray on newspapers so that the excess starch which overflows can be subsequently discarded. The starch solution is allowed to cool slightly (until visible steam no longer arises from the surface) and the lid is placed on the gel tray in a manner to prevent air bubbles from being trapped between the lid and the solution. This is done by starting at one end and placing spring clamps on each side. The top is gently forced into place working from the clamped end. As the top comes into place it squeezes the still liquid starch solution out of the sides and remaining end. The top is clamped all the way around (3 clamps on each side). The template is then placed into the holes of the gel tray lid and held in place with tape (see Figure 3). Excess starch solution from the flask (or that which has run off the gel tray) is used to completely seal the area around the template where it enters the tray lid.

The gel tray is allowed to stand at room temperature for 1 hour and is then refrigerated for approximately 2 hours, after which it is ready for use. If desired, the gel may be kept sealed in the gel tray for several days at refrigerator temperature before use.

At the time of use, all of the clamps except the middle one on each side should be removed; then the two endplates are removed, and, finally, the two middle clamps. A spatula is used to pry the gel lid (with attached template) off of the gel tray. This should be done gently allowing the gel tray lid and gel to separate slowly so that none of the gel sticks to the lid.

2. Short Method

Many types of electrophoretic runs require only 4 hours and can be completed in an 8 hour day if a short method of gel preparation is used. In order to allow for 4 hours of electrophoresis and 2 hours of incubation for staining, it is possible to shorten the gel preparation time to 1½–2 hours so that an entire operation can be carried out in one day. Although routinely it is better to utilize the standard method because gels that have been allowed to mature for a few hours are more reliable, the gel preparatiton method can be shortened as follows.

The gel is made according to the standard method outlined above. However, before the templates and cover are placed on the gel, both are coated with a thin, even layer of Petrolatum. After the gel has cooled for 45 minutes at room temperature, it is placed in a refrigerator (2°–6°C) for approximately 15 minutes. It can then be used. One difficulty often encountered with gels that have not cooled for very long is that they stick to the gel cover or template. This difficulty will usually be overcome by the Petrolatum. However, if time permits, it is usually more reliable to prepare the gels the day before and allow them to mature overnight at 4°C.

D. Adding the Sample

After the gel has cooled and the lid and endplates removed, the previously prepared sample (see Chapter 4) may be applied. A disposable Pasteur pipette with a rubber bulb can be conveniently used to suck up the sample from its test tube and place it in the gel slot. (Of course, a record should be made of which samples go into which slots.) The gel slot should be filled with the sample (0.025–0.100 ml, according to slot volume) so that the side of the slot from which the proteins will migrate is covered with solution. If the slot is only partially filled, there may be distortion of patterns as a result of non-uniform migrattion from the slot.

It is, of course, possible to vary the volume of the slot by utilizing templates of varying thickness. This can also be accomplished by using gel tray lids (and accompanying templates) with slots of varying lengths (see Table I). It is not necessary to put samples in all slots; some may be left unfilled, although it is good practice to fill the empty slots with buffer to avoid distortion.

After the samples have been placed in the slots, the tops of the sample slots are sealed to prevent sample loss. It is convenient to use melted Petrolatum for this purpose, although it should not be too hot when it is applied or it will cause damage to the enzymes in the sample. It is good practice to add a few drops of melted Petrolatum (only a few degrees above melting temperature, which is about 54°C) to the gel surface and allow it to flow over the sample slot. This procedure allows the Petrolatum to cool slightly before it comes in contact with the sample. As the Petrolatum solidifies it seals the sample in the sample slot. The entire gel and gel tray, except for the ends of the gel, are then wrapped in plastic wrapping paper, such as Saran Wrap. This prevents undue drying of the gel by evaporation, and it also helps hold the gel in place in the gel tray.

E. Electrophoretic Setup

1. Bridge Buffer Trays, Electrodes, Wicks, and Stands

A diagram of a setup for vertical electrophoresis is shown in Figure 4 in side and frontal view. A photograph of a setup in a refrigerated display case is shown in Figure 5. Buffer trays may be made of plexiglass and, in general, should be the same length as the width of the gel tray with which they are employed, or should be equal in length to the widest gel trays in use in the laboratory, in which case they can be used interchangeably with all gel trays. Our buffer trays are about 29 cm long, about 7.5 cm in depth, and about 5 cm in width (Figure 4). Note that two buffer trays are used at the top, and two are used at the bottom. One of the trays at the top and one at the bottom are used for electrode immersion and are called electrode trays (Figures 4 and 5). The electrode trays are connected by a filter paper wick to the bridge buffer trays (Figures 4 and 5). The electrodes are not immersed in the bridge buffer trays directly because the pH changes which occur near the electrode might effect the pH in the gel. An alternate method to the use of two separate trays at each end is the use of a partition penetrated by filter paper wicks.

Electrodes which are to be immersed in the two electrode trays can be constructed of plexiglass with platinum wires affixed near one edge and connected to a lead (Figures 4 and 5). The electrodes may be of a width (e.g. 10×10 cm) which will fit into the smallest bridge buffer tray and then be used interchangeably with all sizes.

Filter paper wicks (cut from Whatman No. 3 MM) of the same width as the gel tray and of sufficient length to reach the distance shown in Figure 4 are cut from large sheets. These are used to transmit the current from the electrode tray through the bridge buffer tray to the end of the starch gel (Figure 4). In use, the filter paper wicks are first soaked for a moment in the buffer trays. After that, capillary forces keep them wet.

Vertical electrophoresis requires a stand such as the one shown in Figures 4 and 5. These can be conveniently made of wood or plexiglass. It is probably safer not to make them of substances which conduct electricity. If they are made of sufficient width to accommodate the widest trays which are to be used, they can then be used interchangeably with trays of all widths. Another reason for wide stands is to accommodate two narrow (6–12 slot) gels side by side. Note that in the front view (Figure 4) it would be possible to set up a second gel alongside the first and run both from the same setup.

2. Power Supply

Electrical current may be supplied by any one of a number of commercially available units. In listing certain models we have used we are not implying that they are better than other models. For routine use we employ the Heathkit Regulated High Voltage Power Supply, Model IP-17 (which has replaced IP-32) variable from 0–400 V and capable of continuous delivery of 100 mA at 400 V. It is manufactured by the Heath Co., St. Joseph, Michigan. The Beckman model RD-2 Duostat has also seen wide use (Beckman Instrument Co., Fullerton, California). We have successfully used the E-C Model 545 manufactured by E-C Apparatus Corp., Philadelphia, Pennsylvania, Model

26 / Chapter 2. Starch Gel Electrophoresis

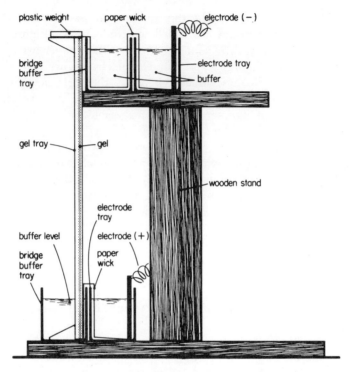

Figure 4. Diagram of a starch gel setup. Above is a side view, and on the facing page a frontal view. See text for explanation.

38201 manufactured by Gelman Instrument Co., Ann Arbor, Michigan, and the Buchler Model No. 3-1008, Buchler Instrument, Inc., Fort Lee, New Jersey. All of these power supplies are capable of producing 500 V. The Model 3-1014A power supply manufactured by Buchler Instruments makes an excellent power supply. It is capable of allowing independent adjustment of voltage and amperage, but is therefore more expensive. It is capable of producing 1000 V. In general, the power supply should be capable of producing at least 12 V/linear centimeter of gel. With a gel 33 cm long, a 400 V power supply will accomplish this.

The power supply should be properly grounded internally. The slightest electrical shock from touching the equipment should immediately cause the unit to be disconnected and repaired. If a worn out wire or faulty connection produces a "short" to the chassis of the unit, it can cause a fatal shock.

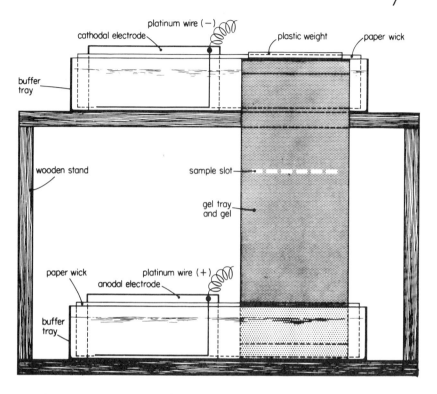

3. Cooling

The gel must be cooled since it tends to warm as a result of the electric current. Undue warming will inactivate enzymes and distort patterns. One approach is to carry out electrophoresis in the cold at slightly above freezing temperatures. A walk-in type coldroom if available, makes an excellent place for electrophoresis. If other work is being carried out in the coldroom, the electrophoretic set up should be sufficiently separated and screened off in some manner so that accidental electrical shocks do not occur. (These are usually not extremely dangerous at the level of current being employed, but could be dangerous under certain conditions.) It is also good practice, at least in our experience, to place the power supply outside of the coldroom. The high humidity which usually exists in a coldroom causes deterioration of the power supply necessitating frequent repairs. It is convenient to have a rack outside the cold room on which one or more power supplies can be placed. Conduit can be led into the coldroom to provide the desired number of outlets. Leads from the platinum buffer tray electrodes

Figure 5. Photograph of a starch gel electrophoretic setup. On the right can be seen several power supplies on a rack. The left two-thirds of the picture shows a part of a refrigerated display case. On the top shelf of the display case is an electrophoretic setup.

can then be plugged into the outlets within the coldroom. The outlets should be clearly marked so that it is apparent from which power supply each lead is derived.

An excellent substitute for a coldroom (and perhaps even an improvement, under some circumstances) is the use of a refrigerated display case, a portion of which is shown in the photograph in Figure 5. (Also shown is the rack for the power supplies.) This particular display case is made by Perlick Refrigeration Co., Milwaukee, Wisconsin, and has worked very well. Such display cases are available in a variety of sizes from a number of companies. The size can be chosen according to the needs of the laboratory. However, some care should be used in specifying the type. Many of the beverage-type display cases used in stores have the disadvantage of requiring a defrosting cycle once every 24 hours, during which they warm up. This type is not appropriate for overnight electrophoresis. Also, the thermostatic controls should be of good quality and allow constant refrigeration down to 1° or 2°C.

There are approaches to cooling the gel during electrophoresis other than the use of coldroom or refrigerator space. Some gel trays are designed such that cold water can be circulated in a jacket on both sides of the gel.

4. Setting Up the Gel

For reasons of safety, the power supply should always be left disconnected while the gel is being set up.

The gel tray is placed in a vertical position facing in with the bottom sitting in the bottom bridge buffer tray (Figures 4 and 5). The gel is usually slanted slightly (Figure 4, side view) or fastened with tape. Usually the shorter end of the gel (starting from the sample slot) is designated the cathodal end and is placed at the top. (If the enzymes under study migrate cathodally, the long end can be made the cathodal end, providing more room for migration.) The bridge buffer and electrode trays are $2/3$–$3/4$ filled with the appropriate buffer (according to the specific method—see Chapter 5). The two wicks are moistened in the buffer trays and placed in position. A piece of plastic about the same dimensions as the end of the gel is usually used to weigh down the wick lying on top of the gel so that it makes solid contact with the gel. The electrodes are placed in the electrode trays (Figures 4 and 5) and plugged into the proper power supply leads. The power supply is turned on and the voltage or amperage adjusted as appropriate.

In general, the power put into the gel system may be regulated by adjusting either the voltage or the current. In many systems in Chapter

5, a general range for both variables is given so that the worker will know whether or not he is obtaining the desired power. The power (amount of work rated in terms of watts) put into the gel is proportional to the voltage × current.

$$\text{Power} = \text{voltage} \times \text{current} = \text{current}^2 \times \text{resistance}$$
$$\text{Voltage} = \text{current} \times \text{resistance}$$

Power is directly proportional to voltage and to the square of the current. The heating of a gel is more markedly affected by increasing current. Thus, in general, buffer systems are designed to attain a relatively high voltage at the expense of current.

More than one gel can be run, hooked up in parallel, from a single power supply. As can be seen from Figure 4, front view, it is possible to set up two small, or one large gel on each side of the stand. Single electrode trays at the top and at the bottom can serve for both sides of the stand. As gels are added, the electrical resistance in the circuit is decreased. Since voltage = resistance × current, the voltage will remain constant and the current will increase as registered on the ampmeter of the power supply. Usually we are interested in either the voltage, or the current per gel. Thus, if a single gel requires 10 mA for adequate migration of isozymes, the power supply should be adjusted to 20 mA for two gels, or else to the specified voltage. Running more than one gel on a setup sometimes results in problems because of dissimilarities in the resistance of the gels; these dissimilarities cause more current to flow through one gel than the other. Also, if resistance changes in a gel due to technical problems it may lead to differences in current through the two gels. For these reasons we have some preference for employing single gels with a large number of slots, rather than multiple gels on one power supply.

F. Electrophoresis

1. General

Once the gel is set up for electrophoresis, and the power supply is turned on and adjusted, there is not much more for the laboratory worker to do until electrophoresis is completed. It is good practice to check the amperage or voltage reading on the power supply after 30–60 minutes to determine that they are holding constant. The temperature in the coldroom or refrigerator can be checked at this time also. Ordinarily, after the gel is set up, it can be ignored until time to remove it. Occasionally the gel will crack and not conduct current well. Little can be done about this besides making a new gel and beginning again. Occasionally the wicks will dry out and current flow will stop. Moistening the wicks will correct this problem.

2. Duration

The purpose of the electrophoretic procedure is to separate the various isozymes of an enzyme system. The duration of electrophoresis is usually determined empirically by the length of time required to obtain adequate migration and separation of the isozymes. This will be influenced by the current flow, the molecular size of the isozymes, their isoelectric points, and the pH and ionic strength of the buffer. The higher the ionic strength of the buffer, the slower the rate of migration, the sharper the bands, and the greater the heating effect. In practice, a compromise is usually reached among these factors. The greater the current employed, the faster the migration rate, although there are constraints on this factor because of gel heating if too much current is used.

In practice the duration of starch gel electrophoresis is usually either 3–5 hours (1 day procedure) or about 18 hours (overnight procedure). There are certain advantages to shorter periods of electrophoresis. If the enzyme is unstable, the shorter the period of electrophoresis, the more activity will remain for staining. Also, if the electrophoretic run is short enough (such as 4 hours), it is possible to make, run, and stain the gels within 1 work day.

G. Slicing the Gel

After completion of electrophoresis the power supply is turned off and the gel tray with the enclosed gel is disconnected from the rest of the electrophoretic apparatus and taken out of the coldroom. The plastic wrapping paper is removed and the triangular end portion of the gel is cut off and discarded. The gel is loosened from the gel tray along the sides by running a spatula along each side. It can usually then be slid out of the gel tray onto a flat plastic or glass surface simply by exerting gentle pressure on one end and on the top surface with the flat part of the hand. If the gel cracks or breaks, it should not be discarded because usable isozyme patterns can be obtained from gels which have broken. Often the gel will break at the site of the slots. This, also, is not particularly harmful.

It is good practice to mark the gels in some manner so that orientation can be maintained regarding the order of the sample slots. There are many ways of doing this. We usually clip off the corner of the gel on the anodal end on the side of slot 1 prior to slicing. If more than one gel is being developed in a day, or if the gel is to be cut into two sections, it is good practice to further mark the gels to indicate their number. We usually use the large end of a Pasteur pipette, or other piece of glass tubing, and punch out one hole in the anodal end of gel 1, 2 holes in the anodal end of gel 2, and so on.

In the case of very wide gels (such as 30 slot gels) the gel may be wider than the available gel slicer and must then be cut longitudinally into two sections prior to slicing. Care must be taken when cutting a gel longitudinally so that the cut is parallel to the edge of the gel and does not run across one or more of the isozyme patterns.

There are several types of gel slicers. We have used two. The first (Figure 6) was made by a local company and consists of a tray not unlike the gel tray. A wire, usually thin piano wire, is stretched under tension across the tray at about the midpoint. The internal width of the tray is equal to or greater than the width of the gel to be sliced. The depth of the tray is critical because the wire is strung across the top of the tray walls, and this determines the thickness of the gel slice. The height of the walls across which the wire is strung must be such that the gel is sliced in the appropriate place. It is possible to

Figure 6. Photograph of one type of gel slicer. See text for description of its use.

G. Slicing the Gel / 33

Figure 7. Photograph of the Buchler gel slicer. See text for description of its use.

construct such a device with interchangeable sides of varying thickness to vary the thickness of the gel slice. The gel is placed on a thin plastic sheet, and the sheet and the gel are pushed through the slicer tray in such a way that the gel is sliced into two horizontal sections. The height of the wire must take into account the thickness of the sheet on which the gel is riding, and the thickness of the gel itself so that the wire hits approximately the midportion of the thickness of the gel. An alternate method of construction of a similar device allows the gel to remain in one place in the tray while the wire, strung on a U-shaped frame, is pushed through the gel (Bloemendal, 1967).

A second type of gel slicer which is considerably better is made by the Buchler Instruments, Inc., Fort Lee, New Jersey. The Buchler gel slicer is shown in Figure 7. It consists of a heavy metal tubular section to which are attached two metal leg devices. Each leg has several slots of varying depths. Piano wire is strung across these slots at the depth appropriate for the thickness of the slice to be made. The design of the Buchler gel slicer is such that it can be made to any width desired. The usual commercial model will accommodate a gel 16.5 cm in width. We have successfully used a modification that will accommodate a 27.5 cm wide gel, and Buchler is willing to make wide slicers upon request. The gel is placed on a flat surface, the slicer is pushed across, with the tubular section over the gel, and with one leg on each side and the wire passing through the gel. This slicer lends itself well to multiple slices of the gel because if the gel is reasonably thick, a cut can be made with the wire at, say, one-third thickness, that slice can be removed, and another cut of the same thickness can be made.

Starch gel must be sliced at least once prior to staining. The outer (uncut) surface does not usually show good banding patterns. However, another reason for slicing gels is to allow multiple enzyme staining from one electrophoretic run; that is one slice can be stained for one enzyme, and another slice stained for a different enzyme, possibly with a protein stain. Obviously the larger the number of slices, the greater the number of studies that can be carried out from one run. However, there is a practical limitation to the number of slices that can be obtained from a gel, because of the friability of the gel if it is too thin. Also, it is not easy to slice a gel uniformly across its entire length, and some areas will be thinner than others. In our experience, 3 slices from a gel of approximately 1 cm thickness is all that can routinely be expected.

H. Staining the Gel

1. Standard Staining Solution Method

After the gel has been sliced it is ready for staining. This consists of placing the slice of gel (cut surface up) in some type of container called the staining dish, to which the staining solution is added until the top surface of the gel is completely covered by the fluid. Pyrex baking dishes purchased in local stores serve quite well as staining dishes. However, there may be some advantage in building staining dishes to specifications out of plastic. This is particularly true if the reagents to be used for staining are expensive. In such a case, the smaller the volume of staining fluid that can be used the less the cost. A staining dish need not be much longer or wider than the gels to be stained.

The reagents in the staining fluid will vary according to the enzyme to be studied (see Chaper 5, Specific Methods). The solution is buffered with a prescribed buffer at a specific pH. The pH must be optimal for the isozymes, or at least provide good enzyme activity. In addition to the buffer, the staining solution will contain the substrate and all necessary cofactors for the enzymes under study. It is the addition of a specific substrate of the enzyme which gives the isozyme technique its specificity. For example, in the glucose-6-phosphate dehydrogenase

H. Staining the Gel / 35

(G-6-PD) isozyme system, the specific substrate glucose-6-phosphate is added. In addition, the necessary cofactor triphosphopyridine nucleotide (TPN) is added.

However, if nothing else were added, the G-6-PD isozymes would not stain. It is necessary to add additional reagents so that a colored product is formed at the site of the enzyme activity being studied. In essence, histochemical techniques are applied to the gel. This may be accomplished by linking one of the reaction products to a dye system that changes from colorless to colored when the reaction product is generated. For example, in the G-6-PD isozyme system, one reaction product is reduced TPN, or TPNH. An electron transport carrier, phenazine methosulfate, and a colorless dye, nitro-blue tetrazolium, are inculded in the staining solution. As the TPNH is formed, it reduces the phenazine, which in turn reduces the nitro-blue tetrazolium. The latter, upon reduction, is converted to an insoluble formazan which is blue. Thus, the G-6-PD isozymes are indicated by a blue color. The formazan precipitate will form only at the point of G-6-PD activity in this system because only the substrate glucose-6-phosphate was added. However, the phenazine–tetrazolium system can be used to indicate the site of any dehydrogenase enzyme which generates TPNH or DPNH, so long as that enzyme's substrate is included in the staining solution.

The above is a typical example of an indicating reaction. There are many types as can be seen from consulting Chapter 5. Occasionally, the reaction product of the enzyme reaction may itself be colored. Many enzymes are stained by so-called nonspecific methods, in which they act on a synthetic substrate which may have little to do with their true physiological role. This group includes the esterases and phosphatases. There are many enzymes which do not as yet have good staining methods available. There is considerable need for further work to develop new histochemical methods for staining enzymes.

Staining is usually carried out for 1–3 hours in a warm-air incubator at 37°C. It is convenient for the incubator to have sufficient racks so that a number of staining dishes can be incubated at one time. Incubation above room temperature at 37°C increases the rate of the enzyme reaction causing the stain to form in a shorter period of time. If staining is too slow, there may be considerable diffusion of the enzyme molecules and less sharpness to the banding patterns.

It is good practice when studying a new system to incubate one slice of the gel with all reagents present in the staining mixture except the substrate of the enzyme. If no bands appear, this helps indicate that the bands which are appearing in the other slices of the gel do, in fact, represent the particular enzyme under study. Of course, other enzyme bands may appear if the substrate is contaminated with materials which may serve as substrate for other enzymes. If this is suspected, additional control studies may be performed in which the coenzyme, specific metal requirements if any, etc., are varied in order to define the isozymes which are appearing. Also, purer substrate can be sought and tested, if available. Some enzymes, catalyzing presumably specific reactions, may on occasion catalyze other reactions which are also thought to be specific. The detection of such cross reactivities usually depend upon staining with multiple substrates.

2. Agar Overlay Method

Often the colored substance generated by an indicating reaction is soluble. If so, and if the indicating reaction is carried out in solution, it will not remain localized at the site of enzyme activity to indicate the site of the isozymes. One way of overcoming this difficulty is the use of an agar overlay. This may be carried out as follows. An appropriate amount of agar is weighed out to provide a 1% solution in the final staining solution. The agar is added to ¾ of the staining solution buffer, and this mixture is heated to boiling. While the agar solution cools, the other reagents of the staining solution are added to the remaining ¼ of the staining solution buffer. After the agar solution cools to 45°C, the ¼ portion of the buffer containing the other reagents is added and the mixture stirred well. The reason for not adding the other reagents prior to the heating step is to prevent their destruction by the high temperatures required to dissolve the agar. The agar mixture is then poured over the surface of the gel in a staining dish until the entire surface is covered. It is allowed to cool to room temperature and to gel. After the agar overlay solidifies, the entire gel in the gel staining dish is incubated at 37°C for as long as required for band development (usually 1–3 hours). The agar gel prevents the reactants from diffusing rapidly away from the site of the enzyme bands. At the time of interpretation of the gel, the agar may be removed if desired. However, the patterns can usually be seen quite well through the agar.

3. Cellulose Acetate or Filter Paper Overlay Method

Recently Knutsen (C. Knutsen, personal communications) has developed a method which serves the same purpose as the agar overlay and conserves expensive reagents. Cellulose acetate strips (or filter paper, which doesn't work quite as well) are soaked in a few ml of the staining mixture and placed on the gel. The bands appear primarily on the overlay, which can be analyzed after removal.

I. Gel Preservation

With some stains it is possible to preserve the starch gel. For example with the stain, nitro-blue tetrazolium, or with most protein stains, the gel can be soaked in 50% ethanol for 30–60 minutes. This causes the gel to become considerably tougher. It can then be wrapped in plastic wrapping paper and preserved in a refrigerator almost indefinitely. Many stains, however, including some of the other tetrazolium dyes, are soluble in alcohol and cannot be preserved in this way.

J. Records

Since many of the staining reactions are transitory and the gels cannot be preserved, it is good practice to develop a system for recording the results promptly. The recording system will vary, of course, according to the type of enzyme systems under study and the needs of the laboratory (see also Chapter 7). If the isozyme banding patterns fall into characteristic types, such as determined by specific genes, it may be sufficient to simply record in a laboratory notebook the designation for those patterns, such as, for example, G-6-PD type A, G-6-PD type AB, or G-6-PD type B. Where the system is less well defined, it may be desirable to have the technician make a sketch of the banding patterns while these patterns are still present on the gel. Perhaps the ideal system is to photograph the gel while the pattern is present, although this is more costly. We have utilized Polaroid cameras for this purpose, and find them very suitable. Pertinent information regarding the date, the system, and the samples can be recorded on the back of the photograph and a permanent record of the actual visual patterns can be maintained. These may be colored, or black and white. We have also used 35 mm colored slides, and these too are quite suitable. It is necessary when using this latter type film to make a label for the gel and include it in the picture since these slides are usually developed in lots of 20 or 36, and the identification of the gels may become quite confusing.

Starch gels can also be scanned, either directly by densitometric methods (see Chapter 7), by densitometric methods after glycerol decolorization (see Section K, this chapter), or by reflectance methods (see Chapter 7).

K. Glycerolization of Gels

If it is desired to quantitate enzyme bands by densitometric methods, starch gels may be rendered transparent by placing them in glycerol warmed to 70°–80°C for 5 minutes. While this procedure will remove the opaque background of the starch gel itself, some residual background stain may persist which can possibly be removed by electrical destaining (see Section L, this chapter).

L. Destaining of Gels

If a gel is suspended between the two electrodes of a power supply (such as those described in Section E, 2, this chapter), such that the flat surface of the gel is perpendicular to the current flux, the background staining (with protein stains) may be removed. This type of destaining may also be used with electrophoretic media other than starch. The basic requirement is that the specifically fixed, precipitated dye not be removed, whereas the dye causing nonspecific background staining must be capable of becoming charged in an electric field and migrating out of the gel. Because of its toughness, Electrostarch is much better for this kind of manipulation than other starches. Commercial destainers for starch gel are sold by Otto Hiller, P. O. Box 1294, Madison, Wisconsin and by E-C Apparatus Corp., Philadelphia, Pennsylvania. Nerenberg (1966) has described destaining techniques in some detail. Most work with destaining has been done with protein stains, and little information is available on the success of this procedure with zymograms.

M. Special Studies

For studying variation in the molecular size of proteins, the starch concentration may be varied from 11.4–15.6% (Smithies, 1962). For subunit analysis of proteins a urea starch gel may be prepared by adding 80–85 gm of hydrolyzed starch powder and 320 gm of urea, to a 2 liter side-arm flask containing 400 ml of buffer. The solution is then heated, degassed, and poured in the usual manner (Poulik, 1960).

Two-dimensional electrophoresis has been carried out with starch gel (Poulik, 1963). Two-dimensional electrophoresis with starch gel as 1 dimension has been used also. Combinations include paper (Smithies and Poulik, 1956), and cellulose acetate (Duke, 1963). Immunodiffusion in thin-layer starch gel has also been employed (Poulik et al., 1959). A bibliography with a large number of references entitled "Starch Gel Electrophoresis" has been compiled by Connaught Medical Research Laboratories, University of Toronto, Toronto, Canada.

REFERENCES

Bloemendal, H. (1967). In "Electrophoresis" (M. Bier, ed.). Academic Press, New York, pp. 379–422.
Boivin, P., Hugou, M., and Hartman, L. (1959). Rev. Franc. Etudes Clin. Biol. 4, 812.
Daams, J. (1963). J. Chromatog. 10, 450.
Duke, E. (1963). Nature 197, 288.
Koch, H., Bergstrom, E., and Evans, J. (1964). Mededel. Koninkl. Vlaam. Acad. Wetenschap. Belg., Kl. Wetenschap. 26, 3.
Latner, A. L., and Skillen, A. W. (1968). "Isoenzymes in Biology and Medicine." Academic Press, New York.
Marsh, C., Jolliff, C., and Payne, L. (1964). Am. J. Clin. Pathol. 41, 217.
Mouray, H., Moretti, J., and Fine, J. (1961). Bull. Soc. Chim. Biol. 43, 993.
Nerenberg, S. T. (1966). "Electrophoresis." Davis, Philadelphia, Pennsylvania.
Poulik, M. (1960). Biochim. Biophys. Acta 44, 390.
Poulik, M. (1963). Nature 197, 752.
Poulik, M., Zuelzer, W., and Meyer, R. (1959). Nature 184, 1800.
Ramsey, H. (1963). Anal. Biochem. 6, 83.
Smithies, O. (1962). Arch. Biochem. Biophys. Suppl. 1, 125.
Smithies, O., and Poulik, M. (1956). Nature 177, 1033.

Chapter 3

Electrophoretic Media Other Than Starch Gel

A. Introduction

The techniques in this chapter are not described in as much detail as the starch gel methods of Chapter 2. Much of the technology for one medium is applicable to others. The power supplies and their use are the same, the buffer and electrode trays may be arranged in similar fashion (or at least the principles are the same), refrigeration methods are similar, etc. The worker using the media described in this chapter should consult Chapter 2 for further details of particular topics.

The gel and bridge buffer systems described in Chapter 5 have been evaluated primarily by starch gel electrophoresis but may be applicable with little or no modification to some of the electrophoretic media described in this chapter. With respect to staining methods, most of the systems described in Chapter 5 will work quite well with media other than starch gel. A number of specific references to isozyme methods using the media under discussion are given in this chapter so that the worker may also try other published systems if he desires.

B. Paper Electrophoresis

Paper electrophoresis was one of the first methods developed for zone electrophoresis (see Chaper 1). In early work its principle use was in the separation of various components of serum proteins in clinical applications. When it became apparent that clinical information could be obtained from the study of multiple molecular forms of serum enzymes, it was only natural to turn initially to paper for the demonstration of these forms. Subsequently paper has been largely replaced, at least for analytical isozyme studies, by other media. It lacks the resolving power of other media; it is opaque, which hampers quantitation; it requires a relatively long period of electrophoresis; proteins become adsorbed, producing "tailing" effects in the patterns; electroosmosis is prominent.

Isozymes which have been studied by the paper technique include the peroxidases (Jermyn and Thomas, 1954), alkaline phosphatases (Baker and Pellegrino, 1954), lactic dehydrogenases (LDH) (Wieland and Pfleiderer, 1957; Sayre and Hill, 1957), and aspartate aminotransferases (Pryse-Davies and Wilkinson, 1958).

Paper electrophoresis can be carried out either horizontally or vertically. With vertical electrophoresis the method consists of hanging a piece (e.g. perhaps 25 cm in length) of Whatman No. 1 to No. 3 MM grade filter paper over a length of nonconducting thread such as nylon. The two ends of the filter paper dip into two bridge buffer trays, which are connected by means of filter paper wicks to two electrode trays, much in the manner explained in Chapter 2, Section E, for the starch gel setup. One electrode tray contains the positive electrode and the other contains the negative electrode. The buffer trays should all be filled to approximately the same level to prevent siphoning. Rather than separate electrode trays, partitions penetrated by suitable wicks may be used. The partitions separate the electrodes from the buffer solution in which the filter paper is immersed. The reason for partitions or separate trays, of course, is to minimize the effect on the electrophoretic media of the pH changes which occur in the buffer near the electrode.

Horizontal paper electrophoresis is similar to vertical with, however, some sort of plastic bridge to provide horizontal support for the paper. The Gelman Instrument Co., Ann Arbor, Michigan manufactures a multipurpose chamber suitable for paper, cellulose acetate, agar, acrylamide, and starch (Figure 8). The Shandon Scientific Co., Ltd., Willesden, London N.W. 10, England supplies an apparatus which is suitable for both paper and cellulose acetate. The E-C Apparatus Corp., Philadelphia, Pennsylvania, also manufactures a "Pressure-Plate Electrophoresis Cell" which is advertised as suitable for paper, agar, starch block, starch gel, and acrylamide gel electrophoresis. The Buchler Instrument, Inc., Fort Lee, New Jersey, also manufactures a migration chamber suitable for paper, cellulose acetate, agar and starch gel.

For sample insertion it is convenient to draw a line across the paper, prior to its becoming wet, with an unsharpened pencil or similar object In vertical electrophoresis this line is usually drawn at the fold. With horizontal electrophoresis it is drawn about one-third of the distance

Figure 8. Photograph of the multipurpose electrophoresis apparatus manufactured by Gelman Instrument Co. Note the two very fine wire electrodes indicated by the arrows. Note that the two compartments are separated by a baffle marked "b." The apparatus has three other incomplete baffles, marked "1," "2," and "3." Paper or cellulose acetate strips can be hung over baffles and 1 and 3, or 2 and 3, depending on their length. Two lengths can be run at one time if desired. Small magnets are used to keep the strips which dip down into the buffer, tight on the troughs. Note that the electrode and baffle arrangement keeps the electrodes a long distance from the ends of the strips. The apparatus has a safety switch so that when the plastic door on top is opened the unit is turned off.

from the cathodal end. The bridge buffer (and electrode trays if any) should be filled with the buffer in use for the particular isozyme system. Then the filter paper should be thoroughly wet with the buffer and equilibrated for a few moments by hanging in the buffer trays prior to sample insertion. It is good practice after everything is prepared to turn on the current for a few minutes before applying the sample and to turn it off again. From 3–12 μl of sample per centimeter of slot length can usually be placed at the sample insertion point, depending upon the thickness of the paper used. The sample slot should not be overloaded because it leads to smearing of the bands. The amount of electroosmosis can be evaluated by also applying a nonionized substance, such as N-dinitrophenylethanolamine, or a nonionized dye which can be easily detected after electrophoresis.

Electrophoresis is carried out in the cold, or with cooling, at approximately 10 V (or less) per linear centimeter of paper. Higher voltages usually lead to heating which may damage the enzyme or distort the patterns due to evaporation of the solvent from the paper. The power supplies described in Chapter 2, Section E are quite suitable for paper electrophoresis. Depending upon the migration of the enzymes to be studied, electrophoresis may be carried out from 4–20 hours. After electrophoresis the isozymes should be stained promptly by the appropriate techniques to prevent diffusion.

C. Agar Gel Electrophoresis

Agar is a polysaccharide, the exact composition of which is unknown, prepared from algae. It gels in concentrations as low as 0.75%. Agar gel is inexpensive, easily handled, and easily preserved. It does not have to be sliced like starch gel prior to staining. Unlike paper, cellulose acetate, and starch gel, it is transparent, a condition which allows direct densitometric measurements after staining. It is intermediate in isozyme resolving power between paper and starch gel, is a very fast method of electrophoresis (25–30 minutes), and was first used for isozyme studies by Wieme (1959a) in studies of LDH and sorbitol dehydrogenase. It has seen widespread use for immunoelectrophoresis.

Isozymes which have been studied by the agar gel technique include, besides LDH and sorbitol dehydrogenase, alkaline phosphatase (Oort and Willighagen, 1961) and malate dehydrogenase (Lowenthal et al., 1961).

Agar preparations apparently vary in composition according to the source and, consequently, electrophoretic migration may vary. Ionagar, of Consolidated Laboratories, Inc., Chicago Heights, Illinois, has seen wide use, as has Special Noble Agar, of Difco Laboratories, Detroit, Michigan.

A 1% solution of agar is made up in suitable buffer. Weime (1959b) employed barbitone buffer at pH 8.4, but any one of a number of buffers may be used (Bodman, 1960). To our knowledge, many of the buffer systems of Chapter 5 have not been tried with agar gel electrophoresis.

The agar solution is heated and allowed to boil until the solution is clear. After the solution has cooled to 45°C, it is poured onto clean microscope slides or other pieces of glass until an even layer about 2 mm thick covers the glass. The glass should be perfectly horizontal so that the gel thickness is uniform. Care should be taken to eliminate air bubbles before the gel sets with a probe or wooden applicator stick. The gel is allowed to stand at room temperature until it sets; this occurs after the gel cools to about 38°C (usually about 1–2 hours).

In some operations it may facilitate gel preparation to prepare a stock 2–4% agar gel (in water) and to remelt and dilute it with buffer prior to use. It should be noted that azide and other antibacterial substances which are sometimes added to agar should not be used in agar gels designed for isozyme studies since these substances may inhibit enzyme activity. Agar gels should be allowed to mature for at least 24 hours prior to use. Gel plates may be preserved almost indefinitely as long as they are sealed to prevent drying and kept refrigerated.

The sample is introduced by making a transverse cut with a razor blade about one-third of the distance from the cathodal end of the slide. If a solution of the sample is to be used, several transverse cuts of 2 mm in length and 1 mm in width may be made; it is best to avoid making the cut so deep that it penetrates through to the glass. Each slot should be sufficient to accommodate approximately 5 μl of sample. If the sample is to be applied by means of filter paper strips (5 $\times$ 2 mm) then the width of the transverse cut may be up to 1.5 cm. In this case the small pieces of filter paper, 2 mm in width, are wet with the sample and placed in the sample slot allowing about 2 mm of space between papers. After 15 minutes most of the sample will have been absorbed from the filter paper, which may then be removed. Care should be taken not to touch the bottom of the slit if it penetrates to the glass, to prevent loss of sample.

It is good practice to turn the power on a few minutes before sample addition to allow buffer equilibration in the gel. The power should be turned off prior to applying the sample. Electrophoresis (usually horizontal) should be carried out in the cold, or with cooling (cooling methods have been reviewed in Latner and Skillen, 1968). The power supplies decribed in Chapter 2, Section E, are quite suitable for agar gel electrophoresis. Filter paper wicks may be used to conduct the current from the buffer trays to the slide. An alternative method using a mold containing agar bridges for conduction of electricity has been described by Blanchaer (1961). More convenient are the versatile units made by Buchler Instrument, Inc., Fort Lee, New Jersey, the E-C Apparatus Corp., Philadelphia, Pennsylvania, and the Gelman Instrument Co., Ann Arbor, Michigan. About 15 mA current per each 1 inch agar slide is applied for 25 minutes. After an electrophoretic run the isozymes should be stained promptly since diffusion occurs in agar gel, although not as rapidly as it does with paper and cellulose acetate.

D. Cellulose Acetate Electrophoresis

Cellulose acetate membranes have several advantages over paper as a media for electrophoresis. The technique is faster (1–2 hours) and has greater resolution, and the strips can be rendered transparent so that bands can be quantitated by direct densitometric scanning. In addition, tailing artifact, a problem with paper electrophoresis due to adsorption of proteins, is eliminated.

Isozymes which have been studied with cellulose acetate electrophoresis include LDH (Wieland et al., 1959), leucine aminopeptidase (Smith and Rutenberg, 1963), creatine phosphokinase (Rosalki, 1965), amylase (Aw, 1966), glucose-6-phosphate dehydrogenase, 6-phosphogluconate dehydrogenase, hexokinase, and diaphorase (the latter four enzymes in unpublished studies, this laboratory).

Cellulose acetate membranes may be obtained from Gelman Instrument Co., Ann Arbor, Michigan from Beckman Products, Fullerton, California and from Consolidated Laboratories, Chicago Heights, Illinois, an outlet for Oxo Ltd., London, England. This last company manufactures a gelatinized cellulose acetate membrane called "Cellogel," which is supplied in the wet state and can be rendered absolutely clear after staining. The Beckman membranes are punched so as to fit the Beckman Microzone electrophoresis system. Cellulose acetate membranes are quite brittle when dry, but become quite pliable and easily handled when wet. By proper treatment, the strips can be rendered transparent. The strips can also be dissolved in organic solvents, facilitating quantitation.

Cellulose acetate membranes are cut into strips (if not already purchased as strips) of appropriate dimensions according to the apparatus to be used. The widths may vary according to the equipment and purposes, but a convenient width is 2.5–5 cm. The Shandon Universal apparatus (Shandon Scientific Co., Ltd., Willesden, London N.W. 10, England) will accommodate 8 membranes 10×2.5 cm each. The apparatus of the Gelman Instrument Co., Ann Arbor, Michigan (Figure 8), has strips that are $1 \times 6¾$ cm or 1×12 cm, and the two sizes can be used simultaneously if desired. Nerenberg (1966) recommends

12 × 5 cm strips for use with the apparatus he has described. Buchler also manufactures a multiple purpose migration chamber suitable for cellulose acetate.

While the strips are still dry, a line is drawn with an unsharpened pencil or similar instrument approximately one-third the distance from the cathodal end of the strip, allowing about 0.5 cm margin on each side of the strip. Any other markings to be made on the strip should also be made before wetting the strips. The strips are then placed on the surface of the buffer to be used to allow removal of air from the pores of the membrane. The disappearance of white opaque areas indicates that the air in the membranes has been replaced with buffer. The choice of buffer will vary according to the electrophoretic method to be used (Chapter 5). The strips are then removed from the buffer and blotted with filter paper to remove excess buffer.

The strips are then placed in the electrophoretic apparatus and should be pulled sufficiently tight so that they do not sag. Commercially available equipment is usually constructed for this. We have previously mentioned the Shandon Universal apparatus, the Beckman Microzone system, the Gelman system (Figure 8), the Buchler chamber, the apparatus described by Nerenberg (1966), and the system manufactured by E-C Apparatus Corp. Several strips can be placed in the apparatus at one time. The appropriate buffer is placed in the buffer trays prior to the addition of the strips, and the appropriate connections are made with filter paper wicks. If comparative studies of isozyme patterns are to be made, it is important to keep the sample application point of the various strips lined up so that the isozyme patterns can be compared. It is good practice to turn the power on a few minutes before sample addition to allow buffer equilibration in the strip. The power should be turned off prior to sample addition. About 5–10 μl of sample is maximal for a 2.5 cm wide strip. The sample groove should not extend closer than about 0.5 cm from the margins of the strip.

The power supplies described in Chapter 2, Section E are quite suitable for cellulose acetate electrophoresis. Approximately 16–20 V per linear centimeter of cellulose acetate membrane is used, or a current of about 2 mA per strip. One to 2 hours of electrophoresis is usually sufficient to obtain good isozyme separation.

Isozyme identification may be carried out by the histochemical procedures described in Chapter 5. Only a small amount of staining solution is required. One method consists of placing a small quantity of the staining solution on a piece of glass, floating the membrane on this solution, and incubating it as 37°C for ½–2 hours, or as long as necessary to complete staining.

If densitometric scanning is desired, cellulose acetate strips can be cleared after staining, depending upon the solubility characteristics of the stain. This is done by washing in 5% acetic acid, treating with 95% ethanol for 1 minute, and immersing in 10% acetic acid and 90% ethanol for 5 minutes. Subsequently the transparent membrane is dried and heated on a glass plate at 60°–70°C for 20 minutes, and then scanned. Full details are given in Nerenberg (1966).

The property of being completely soluble in 10% ethanol and 90% chloroform makes it possible, in some applications, to cut the strips, dissolve them, and quantitate dye or other substances (or activity) with a spectrophotometer.

E. Acrylamide Gel Electrophoresis

1. Introduction

Acrylamide gel and starch gel electrophoresis comprise the two high resolution electrophoretic techniques. They stand alone in terms of their excellence in resolving multiple molecular forms of proteins, such as isozymes. The superior resolving power of these two gel media is due to a sieving effect brought about by the fact that the pore sizes of the gels are of the same order of magnitude as the diameter of the globular proteins migrating through them. In the two sections which follow, the advantages and disadvantages of acrylamide gel versus starch gel as an electrophoretic medium will be considered.

2. Advantages of Acrylamide

As the concentration of acrylamide is varied between 5–30%, the pore size of the gel also varies. (The lower the concentration of acrylamide, the larger the pore size.) This property allows the construction of gels with a varying sieving effect. While it is also possible to obtain this effect with starch gel, the total range of variation in pore size is less than that with acrylamide. This property has been particularly useful in "disc" electrophoresis in which acrylamide gels of several pore sizes are included in one gel tube. Acrylamide gels are not as fragile as starch gel, making them easier to handle. They do not require slicing prior to staining as does starch. A very important advantage is that acrylamide gels are clear, making isozyme bands easier to quantitate by densitometry. Acrylamide gels have relatively better keeping qualities than starch gels, unless the particular staining reaction in use with starch permits alcohol fixation, in which case starch gels can also be preserved very well.

3. Disadvantages of Acrylamide

There are certain disadvantages of acrylamide as compared to starch. The type of acrylamide electrophoresis with the greatest resolving power, disc electrophoresis, having segments of gel with varying pore size, does not lend itself to detailed side by side comparisons. There are always small variations in any electrophoretic run from one gel to another, and the disc method does not permit careful comparative studies or the detection of subtle differences in migration which may arise from genetic variation or other sources. Acrylamide gel, being very "rubbery," does not lend itself at all well (in slabs) to horizontal slicing; this prevents multienzyme studies from one electrophoretic run. Below a pH of 6.5, the gel time is rather long. If a lower pH is desired, it is better to form the gel at a pH of 7.0 and then soak it in the acid buffer overnight. This extra processing for acid systems is a disadvantage. Gelling is affected by exposure to air and, thus, all parts of the gel to be used for electrophoresis must be covered until gelling is complete. Perhaps the most important disadvantage, at least as far as we're concerned, is that using acrylamide gel in accordance with the slab technique (to allow side by side comparisons) at a concentration of 8% acrylamide has not shown as much capacity to resolve isozymes over a wide number of systems as has the use of the starch gel technique. Since the proteins migrate relatively rapidly through acrylamide compared to starch gel under electrophoretic conditions, it may be that the pore size of acrylamide gels during electrophoresis are not as restrictive as starch gel and lead to less molecular sieving.

4. Technique for Forming Acrylamide Slab Gels

The two types of gel preparation will be described separately. For the slab or block type of electrophoresis, the starch gel trays described in Chapter 2 at a length of about 15 cm may be used for molds (see also DeVillez, 1964). More convenient are cells developed by the E-C Apparatus Corp., Philadelphia, Pennsylvania, especially designed for acrylamide gel slabs (Figure 9). Two cells are available, one with a 12 × 17 cm slab area which has slot formers available for 1, 4, 8, and 16 samples, and a survey cell with a 17 × 24 cm slab area which can

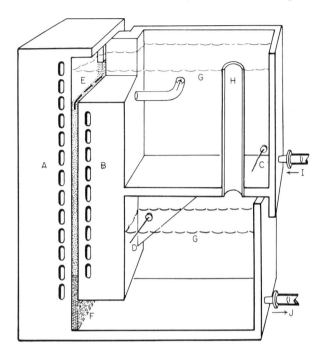

Figure 9. Cross-section diagram of E-C Company vertical gel electrophoresis cell: A, outer cooling plate; B, inner cooling plate; C, upper electrode; D, lower electrode; E, slots for samples; F, sponge strips supporting gel slab; G, levels of buffer in electrode compartments; H, buffer overflow tube; I, buffer in; J, buffer out. Photograph provided by courtesy of E-C Company.

accommodate 16, 24, or 30 samples. Buchler Instruments, Inc., Fort Lee, New Jersey, also makes an acrylamide mold. An apparatus for microelectrophoresis on acrylamide has been described (Pun and Lombroso, 1964).

A 5–30% solution of acrylamide monomer (obtained from American Cyanamid Co., Chicago, Illinois or Eastman Kodak Co., Rochester, New York) is made in the desired buffer which will vary according to the enzyme being studied. (In this connection not all of the buffers developed for starch gel and listed in Chapter 5 have been tried with acrylamide, but most of them should work satisfactorily.) A 0.001 volume of freshly prepared 10% solution of dimethyl aminopropionitrile is then added in the same buffer, followed by 0.01 volume of freshly prepared 10% ammonium persulfate in aqueous solution. After thorough mixing, this solution is poured into a suitable mold and allowed to set for at least 3 hours. The solution is poured into the E-C mold, with the gel compartment horizontal. After the gel has set, the apparatus is turned so that the slab is in a vertical position and filled with buffer solution. It is important that the gel slab be protected from exposure to the air because such exposure prevents polymerization to about a depth of 1 cm. The E-C mold is quite suitable for preventing air exposure.

If starch gel type molds are used, templates such as those described in Chapter 2 can be used to form the slots, as with starch gels. With the E-C molds, slots are also formed by a template device which comes with the equipment. A disadvantage of the E-C mold is the absence of a cathodal section of the gel to observe cathodally migrating isozymes.

It is good practice to turn the power supply on for several minutes before sample addition to allow buffer equilibration in the gel. The power should be turned off prior to sample addition. Five to 50 μl of sample are added, depending upon the size of the slot in use. If the E-C cell is used, addition of the sample is done with a pipette under the buffer. It is good practice to add a few grains of sucrose to increase the density of the sample.

Electrophoresis should be carried out in the cold or by passing cold water through the electrophoresis apparatus. The E-C molds are equipped with water jackets. A voltage potential of about 25–30 V per linear centimeter of gel is applied across the gel for about 2 hours. After electrophoresis, the isozymes are stained by appropriate histochemical techniques, as described in Chapter 5.

5. Technique for Forming Acrylamide Tube Gels

For disc electrophoresis, glass or plastic tubes approximately 7 cm in length and 0.5 cm in internal diameter may be used. They are tightly closed at one end with a rubber stopper. The technique described by Ornstein and Davis (1959) uses a 10–15% (10–15% acrylamide) small pore gel at the bottom of the tube. This is prepared in the same manner as described earlier for the slab technique and is poured into the tubes to within about 1.5 cm of the top. The solution is overlaid with water and allowed to stand 30–45 minutes until polymerization has occurred. Water is removed and a large pore gel (approximately 2.5–5% acrylamide) is poured to within 3 mm of the top. This is overlaid with water and allowed to polymerize. After polymerization of the large pore gel, the gel is ready for use. The mixture to be investigated is gelled in another portion of 2.5% acrylamide, and electrophoresis is initiated.

The large pore gel, often referred to as a spacer gel, serves to concentrate the sample into a narrow starting zone during the electrophoretic run, and the small pore gel actually accomplishes the major separation. Other investigators (Bloemendal, 1967) have eliminated the spacer gel and applied the sample in a buffered mixture with 0.2–0.5 M sucrose.

Of course, many combinations of gels with varying pore sizes may be formed simply by forming a series of gels with various acrylamide concentrations in the tubes. Canal Industrial Corp., Rockville, Maryland, supplies a device designed to permit simultaneous electrophoresis in 8 tubes. Acrylamide electrophoresis in capillary tubes has been described (Grossbach, 1965). Methods for the preparation of acrylamide slab and tube gels, and special applications, are described in some detail by Bloemendal (1967), Nerenberg (1966), Latner and Skillen (1968), Wilkinson (1965), and Hansl (1964). Of particular interest is two-dimensional electrophoresis using gels of two acrylamide concentrations, which yields information on molecular size (Raymond and Aurell, 1962). According to Bloemendal (1967), abstracting services for new reports using this medium are available from Information Center, Canal Industrial Corp., Rockville, Maryland.

F. Choice of Media

With the techniques described in Chapters 2 and 3 the investigator has a reasonable amount of flexibility in setting up his electrophoretic program. For other than strictly routine applications, one of the high resolution techniques, either starch gel or acrylamide gel, should be the primary method. Which of these two is selected will depend upon the aims and needs of the laboratory. If multienzyme staining of a single electrophoretic run with many side by side comparisons is desired, starch is superior. If obtaining the answer rapidly with routine densitometric evaluation is desired, acrylamide is superior.

Besides the selection of either starch gel or acrylamide as the high resolution technique, it may be desirable to complement the program with one of the other techniques, particularly agar gel or cellulose acetate. Once an isozyme pattern is well defined, including genetic variation in a population, and if the lower resolution techniques resolve the isozymes of interest, these techniques can be used quite satisfactorily, and probably more efficiently.

REFERENCES

Aw, S. E. (1966). *Nature* **209**, 298.
Baker, R. W., and Pellegrino, C. (1954). *Scand. J. Clin. Lab. Invest.* **6**, 94.
Blanchaer, M. C. (1961). *Clin. Chim. Acta* **6**, 272.
Bloemendal, A. (1967). In "Electrophoresis" (M. Bier, ed.). Academic Press, New York, pp. 379–422.
Bodman, J. (1960). In "Chromatographic and Electrophoretic Techniques. Vol. II. Zone Electrophoresis" (I. Smith, ed.), p. 91. Heinemann, London.
DeVillez, E. J. (1964). *Anal. Biochem.* **9**, 485.
Grossbach, U. (1965). *Biochim. Biophys. Acta* **107**, 180.
Hansl, R., Jr. (1964). *Ann. N. Y. Acad. Sci.* **121**, 391.
Jermyn, M. A., and Thomas, R. (1954). *Biochem. J.* **56**, 631.
Latner, A. L., and Skillen, A. W. (1968). "Isoenzymes in Biology and Medicine." Academic Press, New York.
Lowenthal, A., Van Sande, M., and Karcher, D. (1961). *Ann. N. Y. Acad. Sci.* **94**, 988.
Nerenberg, S. T. (1966). "Electrophoresis," Davis, Philadelphia, Pennsylvania.
Oort, J., and Willighagen, R. G. J. (1961). *Nature* **190**, 642.
Ornstein, L., and Davis, B. J. (1959). "Disc Electrophoresis." Distillation Products Industries (Division of Eastman Kodak Co.).
Pryse-Davies, J., and Wilkinson, J. H. (1958). *Lancet* **i**, 1249.
Pun, J. Y., and Lombroso, L. (1964). *Anal. Biochem.* **9**, 9.
Raymond, S., and Aurell, B. (1962). *Science* **138**, 152.
Rosalki, S. B. (1965). *Nature* **207**, 414.
Sayre, F. W., and Hill, B. R. (1957). *Proc. Soc. Exptl. Biol. Med.* **96**, 695.
Smith, E. E., and Rutenberg, A. M. (1963). *Nature* **197**, 800.
Wieland, T., and Pfleiderer, G. (1957). *Biochem. Z.* **329**, 112.
Wieland, T., Pfleiderer, G., and Ortanderl, F. (1959). *Biochem. Z.* **331**, 103.
Wieme, R. J. (1959a). "Studies on Agar-Gel Electrophoresis." Arscia, Brussels.
Wieme, R. J. (1959b). *Clin. Chim. Acta* **4**, 317.
Wilkinson, J. H. (1965). "Isoenzymes." Lippincott, Philadelphia, Pennsylvania.

Chapter 4
Sample Selection and Preparation

A. Introduction

Each laboratory will wish to adopt its own modifications of sample preparation methods to suit its purposes. Laboratories interested in screening will utilize faster but less precise approaches to sample preparation, in contrast to a more careful approach by those laboratories in which the specific characteristics of an isozyme pattern in a given organism are under study. The methods presented in this chapter are general guides to the preparation of many different types of samples.

B. Specific Procedures

1. Mammalian Erythrocytes

Erythrocytes which have been studied in our laboratory by the following procedures include human and a number of subhuman primate species, as well as sheep, goats, cattle, swine, dogs, rats, and mice.

a. ANTICOAGULANT

Blood is usually anticoagulated for electrophoretic studies of erythrocyte enzymes, although this is not always necessary (see Section B, 1, b, this chapter). Any one of several anticoagulants may be used including heparin, ethylenediaminetetraacetic acid (EDTA), and acid–citrate–dextrose (ACD). Formula A ACD contains 0.8 gm of citric acid, 2.2 gm of trisodium citrate, and 2.45 gm of dextrose per 100 ml of solution. The addition of 0.15 ml of ACD per 1.0 ml of blood provides good anticoagulation and cell preservation. We usually use ACD or add it after collection in other anticoagulants because of its superior storage properties. Anticoagulants which are metabolic inhibitors, such as fluoride and oxalate, should not be used unless their effects on the enzymes of interest are studied first. Blood may be drawn into either syringes or vacuum tubes (which can be purchased containing the proper amount of ACD, heparin, or EDTA, from Becton-Dickinson Co., Rutherford, New Jersey).

b. Blood Storage

Blood samples may be stored in a number of ways. The length of time during which the sample may be stored will vary with the stability of the enzyme and the conditions of storage. As a general method of storage for at least 1 month and, considerably longer for most enzymes, addition of ACD solution in a ratio of 0.15 ml of ACD to 1 ml of blood is generally satisfactory. This amount of ACD may be added to blood even though it has initially been drawn in heparin or EDTA. The benefits of ACD solution are that it provides an adequate amount of dextrose for metabolism, and that it lowers the pH to about 7.0, decreasing metabolism to a minimal level. The blood is then stored at refrigerator temperatures ($+2°$ to $+6°C$).

Many enzymes will be adequately preserved if hemolysates are stored frozen at $-20°$ to $-70°C$. However, some will not. In general, the colder the storage temperature, the better the preservation of hemolysate enzymes.

Recently, we have experimented with clotted blood stored for over a year at $-70°C$. At the time of usage, the samples were thawed, and the clots broken up with a wooden applicator stick. After centrifugation, the supernatant hemolysate was used. It was possible to type most of the red cell enzyme system studied. Those typed included glucose-6-phosphate dehydrogenase, 6-phosphogluconate dehydrogenase, acid phosphatase, adenylate kinase, esterase, malic dehydrogenase, catalase, lactate dehydrogenase, and phosphoglucomutase.

An alternative method which gives the best storage available, but is rather tedious and time consuming, involves glycerolization of red cells and their storage in the liquid state for long periods of time (up to several years) in liquid nitrogen freezers. The glycerol solution consists of 4 volumes of glycerol and 6 volumes of a solution containing 3.25% tripotassium citrate, 0.6% K_2HPO_4, and 0.4% KH_2PO_4. The glycerol solution is mixed slowly in equal volume with packed red cells (plasma removed). The addition must be slow and accompanied by mixing, or the red cells will be damaged. At the time of usage, the sample is ordinarily deglycerolyzed, a rather tedious procedure. The red cells are washed five times with solutions of decreasing glycerol content (16, 8, 4, 2, and 0%) in saline. It is possible to study the following isozyme systems without deglycerolizing the sample: adenylate kinase, 6-phosphogluconate dehydrogenase, phosphoglucomutase, and acid phosphatase (Weitkamp et al., 1969).

c. HEMOLYSATE PREPARATION

A convenient amount of whole blood (0.5–2.0 ml) is centrifuged at approximately 1000g for 5 minutes. With most enzymes it is not necessary to carry out these steps in the cold. However, some mutants, such as G-6-PD variants, may be quite unstable and retain activity better if handled in the cold. The plasma and buffy layer (leukocytes and platelets) are aspirated and discarded. Although about 10–20% of the leukocytes will remain, these cells are in such a minority that they do not ordinarily show banding patterns which would confuse the interpretation of the erythrocyte patterns. The red cells are washed by the addition of 1 volume of 0.85% saline and the centrifugation step is repeated. After aspiration of the saline, a similar washing is carried out two additional times. After the last aspiration of saline, 1 volume of distilled water is added to the packed red cells, followed by 0.4 volume of toluene. This mixture is shaken well for approximately 20–30 minutes with a shaking device and is then centrifuged at approximately 2000g for 20 minutes. The toluene and lipid layer at the top of the hemolysate is aspirated and discarded. The remaining hemolysate is centrifuged at approximately 10,000g for 20 minutes. The clear supernatant is removed from the button of cell stroma at the bottom and is used for electrophoresis.

The toluene is added in this method to get rid of lipids and lipoproteins. This leads to clarification of banding patterns by minimizing streaking. In our experience the toluene step does not damage enzymes, nor change the isozyme patterns. It is not, however, essential to include the toluene extraction step. Similar remarks may be directed toward the high speed centrifugation at the end of the preparatory procedure. This leads to removal of the red cell stroma and the particulate matter which tends to cause streaking in the pattern. It is not, however, an essential step if a high speed centrifuge is not available.

2. Mammalian Leukocytes and Platelets

In contrast to the situation with erythrocytes, blood from which leukocytes and/or platelets are to be separated should always be used in the freshest possible condition. Leukocytes and platelets begin to clump together and stick to surfaces after the blood is drawn so that yield is best with very fresh blood. Because of the tendency of these formed elements to stick to surfaces, it is good practice to siliconize equipment which comes in contact with the blood or to use plastic non-wettable equipment.

Heparin, ACD, or EDTA may be used as anticoagulants. Platelets may be obtained by centrifugation of 10–20 ml of whole blood at 500g for 5 minutes. This amount of centrifugation will remove the erythrocytes and leukocytes from the supernatant, but leave the platelets still in suspension. The supernatant can be removed and centrifuged at high speed (2000–20,000g) for 10–20 minutes (depending upon speed of centrifugation). Platelets are sedimented as a button at the bottom of the tube. The platelets may be disrupted by sonication, by repeated freezing and thawing, or by addition of a detergent such as Triton-x (obtained from Rohm and Haas, Philadelphia, Pennsylvania. A tenfold dilution of the Triton-x is made, and 1 volume is used per ¼–1 volume of packed platelets). After disruption, high speed centrifugation (e.g. 20,000g) will help clear the supernatant of particles and minimize streaking on the gel. The supernatant is used for electrophoresis.

If the leukocytes of the sample are to be separated, the original platelet-free supernatant obtained after the first 500g centrifugation is added back to the packed cells. After reconstitution of the whole blood sample with the platelet-free plasma, the leukocytes may be separated by addition of sufficient dextran (average molecular weight about 235,000) to make a 3% solution in the blood sample (10–20 ml of reconstituted whole blood is a convenient amount for leukocyte separation). Dextran causes clumping of the erythrocytes which then settle more rapidly than the leukocytes. The mixture is allowed to stand undisturbed in the test tube for 30 minutes. Then the supernatant which contains the leukocytes is removed. If desired, a leukocyte cell count may be made on the supernant using standard hematological leukocyte counting methods. The leukocyte suspension is then centrifuged, the supernatant is discarded, and the button of leukocytes at the bottom is disrupted by sonication, repeated freezing and thawing, or the addition of a detergent such as Triton-x. (A tenfold dilution of the Triton-x is made and 1 volume used per ¼–1 volume of packed cells.)

3. Cell Cultures

Obtaining material from mammalian diploid cell cultures for electrophoresis is much the same as obtaining cellular elements from blood.

The cells from culture are packed by centrifugation, and then disrupted by sonication or addition of Triton-x. (A tenfold dilution of the Triton-x is made, and 1 volume of this used per ¼–1 volume of packed cells.) For glucose-6-phosphate dehydrogenase studies, the Triton-x method is superior to sonication (G. Darlington, personal communication). A lesser amount of cells is required and streaking is less frequent. A high speed centrifugation (18,000g) after the disruption step is also important in eliminating streaking. Triton-x disruption has also been used for studies of 6-phosphogluconate dehydrogenase, lactate dehydrogenase, malate dehydrogenase, and adenylate kinase with acrylamide electrophoresis, and produces equivalent results to sonication or freezing and thawing (A. Velazquez, personal communication).

4. Serum

The use of isozyme techniques applied to serum has considerable potential in clinical medicine (see Chapter 6). For example, it is often of assistance in the diagnosis of myocardial or hepatic disease to know the type of lactic dehydrogenase isozyme patterns which are present in the serum. The general principle behind this application is that tissues vary in their composition of isozymes and tissue damage causes the release of the isozymes of the damaged tissue into the serum. This application of isozymology appears to be in its infancy. Only a few of the many available isozyme systems seem to have been studied so far in application to a wide variety of clinical disorders. By far, the largest number of studies have been carried out with lactic dehydrogenase. There is every reason to anticipate that the future will bring increasing application of isozyme techniques to this area of medicine (see Chapter 6).

Since, for the most part, the isozymes of clinical interest in human serum do not occur except in pathological conditions, considerable attention should be paid to performing the electrophoretic run with fresh material and in a manner which allows maximal activity of the isozymes to be demonstrated. Thus, it is critical that serum be very fresh, that heating during electrophoresis be prevented, and that the electrophoretic run be as short as possible. In this connection, acrylamide gel, agar gel, and cellulose acetate have advantages over starch gel, or filter paper electrophoresis. So far, most implications have been drawn from the quantity of isozymes present (such as, for example, an increase in serum LDH-1 after coronary occlusion) rather than from qualitative differences. Thus, quantitation is important in this application, and here too the transparency of acrylamide gel and agar gel give these media an advantage over the other media.

To obtain serum for electrophoretic study, fresh whole blood is drawn without anticoagulant and placed in a glass tube. As soon as the clot forms the sample is centrifuged and the supernatant serum is used for electrophoresis. It is possible, of course, to concentrate the proteins in serum if the isozyme patterns of trace amounts of protein are to be investigated. Such concentration can be accomplished by dialysis, by pressure filtration, and by a number of other procedures. It is important to know the effects of such procedures on the activity of the isozymes being studied. If the enzymes are very unstable, delays inherent in serum concentration may not be tolerable unless the enzymes can be stabilized in some manner.

5. Animal Soft Tissues

a. OBTAINING THE SAMPLE

Isozyme studies of almost any soft tissue can be performed. To render interpretation most straightforward, certain precautions should be observed. The tissue sample should be obtained as fresh as possible. Postmortem changes should be minimized by refrigeration of the body or organism until the sample can be obtained, since alterations in the isozyme patterns may occur otherwise. Of course, it is unacceptable to collect samples after procedures such as formalin fixation. The tissue sample should be washed as free of blood as possible, otherwise isozymes of blood component may confuse the findings regarding the tissue isozymes. This can be partially accomplished by blotting the tissue sample on filter paper immediately after its removal from the body. With some organs, removal of blood can be accomplished by perfusion of the vascular system with a solution of normal saline.

b. TISSUE STORAGE

In general, the colder the storage, the better the chance of preserving the isozyme patterns. For very long-term storage, a liquid nitrogen freezer is ideal. We have obtained good results with most enzymes by storage of tissues at $-70°C$. Many enzymes will be adequately preserved for a period of several months, or even a year or longer, with storage at $-20°C$. It has been our experience that it is better to store the tissue as a block rather than after homogenization.

c. Tissue Preparation

If it is available, about 1.0 gm of the tissue is placed in a Virtis (or similar homogenizer) grinder cup. (If only small amounts are available, a few milligrams of tissue may be studied, provided that the isozyme activities are strong enough. In this case, homogenization of the tissue may be better carried out by methods described under Sections B, 6, 7, or 8 of this chapter.) Approximately an equal volume (1.0 ml) of buffered saline, pH 7.4, is added. The tissue is ground for 1 minute or until it is relatively well homogenized. The contents of the cup are then centrifuged at approximately 10,000g for 10 minutes. The supernatant material is used for electrophoresis.

The high speed centrifugation accomplishes removal of the particulate matter. This tends to minimize streaking on the gel. Reasonable results can also be obtained by low speed centrifugation if a high speed centrifuge is not available. Unless the stability characteristics of the enzyme under investigation are known, it is best to prepare a fresh homogenate each time an electrophoretic run is to be performed.

6. Drosophila

The following methods may also be used for small insects and organisms of a size similar to *Drosophila*. The flies are killed with ether that is allowed to dissipate before the flies are used. With all of the enzymes of *Drosophila* we have studied, it is possible to use individual flies for electrophoretic studies. This adds great strength to the method because it allows determination of genetic variation between flies. One fly is placed in a 5 ml plastic centrifuge tube, and a tiny drop (from a Pasteur pipette) of the buffer to be used in gel preparation (gel buffer) is added. The fly is ground in the buffer with a glass stirring rod. After maceration of the fly is accomplished, an additional 0.1 ml of buffer is added. This material may be centrifuged, or the larger particles may be allowed to settle. Usually, there is insufficient material for centrifugation. A small amount of the supernatant is used for electrophoresis.

In a similar manner it is possible to study the individual larvae or pupae of *Drosophila*. It is also possible to study the tissue isozyme patterns of pooled tissues of *Drosphila* after dissection under a dissecting microscope. Without a great deal of trouble one can obtain isozyme patterns from the digestive organs, wing muscle, testis, and ovary. For most of these tissues, it is necessary to pool the organs from more than one fly, or the isozymes will be too weak to identify after staining.

We have always used *Drosophila* immediately after killing the flies. Presumably it would be possible to store flies frozen if desired for the subsequent study of many enzymes.

7. Wheat Seeds

This method is also applicable to other seeds of the same general nature and size. The wheat seeds are soaked in distilled water for 1 day. Alternatively, they may be placed between two layers of filter paper in a petri dish and soaked with distilled water for 1 day. Approximately 15 minutes prior to use, the wheat seeds are placed in the gel buffer to be used in the electrophoretic system and allowed to soak. An individual wheat seed is placed in a 5 ml plastic centrifuge tube and 0.1 ml of the gel buffer is added. The seed is crushed with a glass stirring rod and then ground until the larger particles have been smashed. This material is then centrifuged, or the larger particles are allowed to settle. The supernatant is used for electrophoresis.

Soaking the seed for 24 hours initiates germination which causes many enzymes to appear. It also softens the seed and makes it easier to grind. The seeds can be used dry without any prior soaking, however. Since germination begins when soaking is initiated, seeds may be allowed to soak as long as desired and ontogenetic variation can be studied. Many enzymes will show variation according to the period of germination. This means, of course, that if between-seed variation is being studied (rather than ontogenetic differences) the period, as well as the temperature of soaking should be held constant.

8. Plant Tissues

This method may be used for the study of leaves, roots, flowers, stems, and other plant tissues. We have used it with the following plants: *Marchantia* (liverwort), *Funeria* (moss), *Lycopodium* (club moss), *Selaginella* (club moss), *Cycas*, *Psilotum* (whiskfern), *Lygodium* (primitive fern), *Pilularia* (advanced fern), *Podocarpus* (conifer), *Degenaria*, *Urginea* (lily), corn, wheat, orchid, and *Franseria* (composite).

Approximately 120 mg of the plant tissue is ground with purified sea sand in a small mortar and pestle to which is added 0.2 ml of the gel buffer. Grinding should be continued until the tissue is well macerated and the mixture is homogeneous. This mixture is then transferred to a 10×80 mm plastic centrifuge tube and is centrifuged at approximately 10,000g for 6 minutes. The resulting supernatant is used for electrophoretic study.

With simpler plants it is necessary to grind the whole organism. In the cases of plants such as wheat and corn, if leaves are to be studied, hypocotyls from week-old seedlings which have been germinated in petri dishes may be used.

Extracts may also be prepared in a tissue homogenizer, rather than a mortar and pestle, using the same tissue buffer ratio as with the mortar and pestle method. The enzyme systems we have studied have not shown differences between samples prepared by the two different procedures.

It should be noted that particularly in comparisons between species, there may be great differences in protein concentrations between equivalent tissues of the two species. A given species may have much more cellulose in its leaves, for example. One approach to this problem is to prepare extracts which have equivalent protein concentrations by appropriate dilution of the more concentrated extracts.

9. Microorganisms

Microorganisms (bacteria, algae, molds, etc.) for study may be obtained in a number of ways. They can be scraped from culture plates or grown in broth and recovered by centrifugation. If desired, the amount of organisms being used can be standardized by measuring protein concentration. In any case, 100–500 mg of microorganism-rich material can be suspended in a small amount (1–2 ml) of gel buffer, and the microorganisms can be disrupted by sonification. If a sonicator is not available, a reasonable amount of cellular disruption can be accomplished by grinding with sand in a mortar and pestle. Cellular debris may be allowed to settle or may be centrifuged, and the supernatant may be used for electrophoretic study.

REFERENCE

Weitkamp, L. R., Sing, C. F., Shreffler, D. C., and Guttormsen, S. A. (1969). *Am. J. Hum. Genet.* (in press).

Chapter 5

Specific Electrophoretic Systems

A. Introduction

Section E of this chapter describes a large number of electrophoretic systems, all of which have been used in our laboratory. In many cases these methods are modifications of methods in the literature. Obviously, as time goes by, many workers and laboratories will contribute to the evolution of methods such as these. In presenting the methods as we use them we are not implying that they are superior to alternate methods, but simply that these methods here have worked successfully for us in certain applications. Some work has been done to modify methods wherever possible so that they use identical buffer systems and permit multienzyme staining of separate slices. The order of presentation of methods in this chapter is primarily oriented so that methods using similar buffer systems are grouped together, rather than organized according to metabolic relationships. Even though the buffers and/or other parts of the method are often similar from one system to another, they are listed in detail with each method. In this way, each method is reasonably complete as described. Most of the systems have been evaluated primarily with starch gel as the medium, but the basic methods should work as well with most of the other media. Note that Table II lists the various organisms and tissues which have been studied by the methods described.

Section F of this chapter briefly lists and describes certain electrophoretic systems described in the literature which have not been set up in our laboratory. We have arbitrarily divided methods into Section E and Section F on the basis of personal use. This does not imply any basic differences in the methods in the two sections, other than a greater knowledge on the part of the author of the methods in Section E.

B. General Comments about Buffer Systems

In general, buffer systems used in electrophoresis may be viewed as continuous or discontinuous. Continuous buffer systems use the same buffer in the bridge buffer trays and in the gel. Usually the gel buffer is one-tenth the strength of the buffer in the tray. Discontinuous buffer systems use different buffers in the trays and in the gel, and in this case also the gel buffer is usually considerably weaker than the bridge buffer. An occasional system employs 10% sodium chloride in the electrode tray. This will be pointed out where applicable.

As pointed out in Chapter 1, Section C, the type, pH, and ionic strength of the gel buffer will have marked influences on the system. Buffers of low ionic strength permit fast migration and cause relatively little heat development, while buffers of higher ionic strength give sharper zones. The net charge on a given protein molecule with a given isoelectric point will be determined by the pH of the buffer; net charge, of course, influence migration. Beyond these effects, the type of buffer will also influence migration, separation, sharpness of bands, etc. Most buffer systems have been developed for particular enzyme systems empirically and can often be improved by further experimentation.

C. Staining Systems

1. General

In order to identify isozymes, it is necessary to stain them by various techniques. Many different approaches have been used, mostly borrowed from the much older field of histochemistry. Work has been limited almost exclusively, at least in zone electrophoresis, to visual identification. The product of the enzyme reaction, while it does not usually have a color itself, may cause some other substance to be colored, or be linked to a color producing system (for example, see the discussion of the tetrazolium system in Section C, 2, this chapter). If the colored end product is a precipitate or does not diffuse rapidly, such a reaction can be conveniently carried out in solution. If the end product is soluble, or if the formation of color involves a complex coupling process with many soluble intermediates, an agar overlay (see Chapter 2, Section H, 2) may be used to hold the reactants and/or the color in the area of the enzyme molecule.

Table II
PRESENCE OF ENZYME BANDS IN VARIOUS TISSUES AND ORGANISMS[a]

	Glucose-6-phosphate dehydrogenase	6-Phosphogluconate dehydrogenase	Phosphohexose isomerase	Glucokinase	Fructokinase	Phosphoglucomutase	Achromatic regions	Lactate dehydrogenase	Esterase	Carbonic anhydrase	Adenylate kinase	Alkaline phosphatase	α-Glycerophosphate: dehydrogenase	Glyceraldehyde-3-phosphate dehydrogenase	Isocitrate dehydrogenase
Mammalian tissues															
Erythrocytes	+	+	+	+	N	+	+	+	+	+	+	0	0	+	+
Sera	N	N	N	N	N	N	N	+	N	N	N	N	N	N	N
Leukocytes	+	+	N	+	N	N	N	N	N	N	N	N	N	N	N
Liver	+	N	N	+	N	N	+	N	N	N	N	N	+	N	N
Heart	+	N	N	+	N	N	+	N	N	N	N	N	+	N	N
Skeletal muscle	N	N	N	N	N	N	N	N	N	N	N	N	N	N	N
Drosophila															
robusta	+*	+*	+	+*	+	+	+	N	+	N	+	+	+	+	+
willistoni	+*	+*	N	+*	N	+	+	N	+	N	+	+	+	N	+
prosaltans	+*	+*	N	+*	N	+	+	N	+	N	+	+	+	N	0
melanogaster	+*	+*	N	+*	N	+	+	N	+	N	+	+	+	N	+
Bacteria															
S. typhimurium	+*	+	N	+	N	+	+	0	0	N	+	0	0	0	0
E. coli	+*	+	N	+	N	+	+	+	0	N	+	+	0	+	+
Plant genuses (whole plant)															
Chlamydamonas	+*	+	N	0	N	0	0	0	0	N	0	0	0	N	0
Zygnema	0*	0	N	0	N	+	0	0	0	N	+	+	0	0	0
Chara	0*	+	N	+	N	+	0	0	0	N	+	+	0	N	0
Neurospora	+*	+	N	+	N	+	+	+	+	N	+	+	0	+	+
Marchantia	0*	0	N	+	N	+	+	0	+	N	0	+	0	N	0
Selaginella	+*	+	N	+	N	+	+	0	+	N	+	+	0	N	0
Funeria	0*	+	N	+	N	+	+	0	+	N	+	+	0	0	0
Pilularia	0*	+	N	0	N	0	+	0	+	N	0	+	0	0	0
Plant genuses (leaves)															
Cycas	+*	+	N	+	N	+	+	+	+	N	+	+	0	N	0
Psilotum	+*	+	N	+	N	+	+	+	+	N	+	+	0	N	0
Lycopodium	+*	+	N	+	N	+	0	0	+	N	+	+	0	+	0
Lygodium	+*	+	N	+	N	0	0	0	+	N	+	+	0	+	0
Podocarpus	0*	0	N	+	N	+	+	0	+	N	+	+	0	+	0
Degenaria	0*	0	N	+	N	+	+	0	+	N	+	+	0	+	0
Ugineae	+*	+	N	+	N	+	0	+	+	N	+	+	0	+	0
Oncidium	0*	+	N	+	N	+	0	+	+	N	+	+	0	+	0
Francinaria	0*	0	N	+	N	0	+	0	+	N	+	+	0	+	0
Zea mays	+*	+	N	+	N	+	+	0	+	N	+	+	0	N	0
Triticum and *Aegilops* species	+*	+*	N	+	N	+	+	N	+	N	+	+	+	+	+
Plant genuses (seeds)															
Triticum and *Aegilops* species	+*	+*	+	+	N	+	+	N	+	N	+	+	+	+	+
Commercial enzyme	N	N	N	N	N	N	N	N	N	N	N	N	N	N	N

[a] Key to symbols: +, bands present; 0, no bands seen; N, not studied; *, alternate method (see text); **, patterns identical with adenylate kinase patterns.

Often the end product of an enzyme activity may be used as the substrate of another enzyme which can generate a colored product. This is called "coupling." For example, in the fourth system described in this chapter (glucokinase, Section E, 4), the staining reaction involves

Succinate dehydrogenase	Leucine amino-peptidase	Peroxidase	Fumarase	Aldolase	Pyruvic kinase	Creatine kinase	Catalase	Glutathione reductase	Malate dehydrogenase	Alcohol dehydrogenase	Diaphorase	Triosephosphate isomerase	Acetylcholinesterase	Pseudocholinesterase	Glutamate dehydrogenase	Xanthine dehydrogenase	Acid phosphatase	ATPase
0	+*	0	0	+	+**	+**	+	+	+	0	+	N	+	0	+	N	+*	N
N	N	N	N	N	N	N	N	N	N	N	N	N	0	+	N	N	N	N
N	N	N	N	N	N	N	N	N	N	N	N	N	N	N	N	N	N	N
+	N	N	+	N	+**	+**	N	N	N	+	N	+	N	N	+	N	N	N
+	N	N	+	N	+**	+**	N	N	N	N	N	+	N	N	+	N	N	N
N	N	N	N	N	N	+	N	N	N	N	N	+	N	+	N	N	N	N
N	+	N	+	N	+	+**	N	N	+	+	N	+	N	N	+	+	N	+
0	+	N	N	N	N	N	N	N	+	+	N	N	N	N	N	N	N	N
0	+	N	N	N	N	N	N	N	+	+	N	N	N	N	N	N	N	N
0	+	N	N	N	N	N	N	N	+	+	N	N	N	N	N	N	N	N
N	0	N	N	N	N	N	N	N	+	N	N	N	N	N	N	N	N	N
N	+	N	N	N	N	N	N	N	+	+	N	N	N	N	N	N	N	N
N	0	N	N	N	N	N	N	N	+	N	N	N	N	N	N	N	0	N
N	+	N	N	N	N	N	N	N	+	N	N	N	N	N	N	N	0	N
N	+	N	N	N	N	N	N	N	+	0	N	N	N	N	N	N	0	N
N	+	N	N	N	N	N	N	N	+	+	N	N	N	N	N	N	0	N
N	+	N	N	N	N	N	N	N	+	N	N	N	N	N	N	N	0	N
N	+	N	N	N	N	N	N	N	+	N	N	N	N	N	N	N	+	N
N	+	N	N	N	N	N	N	N	+	0	N	N	N	N	N	N	0	N
N	+	N	N	N	N	N	N	N	0	N	N	N	N	N	N	N	0	N
N	+	N	N	N	N	N	N	N	+	+	N	N	N	N	N	N	0	N
N	+	N	N	N	N	N	N	N	+	N	N	N	N	N	N	N	0	N
N	+	N	N	N	N	N	N	N	+	+	N	N	N	N	N	N	+	N
N	+	N	N	N	N	N	N	N	+	+	N	N	N	N	N	N	+	N
N	+	N	N	N	N	N	N	N	0	+	N	N	N	N	N	N	+	N
N	+	N	N	N	N	N	N	N	+	+	N	N	N	N	N	N	+	N
N	+	N	N	N	N	N	N	N	+	0	N	N	N	N	N	N	+	N
N	0	N	N	N	N	N	N	N	+	+	N	N	N	N	N	N	0	N
N	+	N	N	N	N	N	N	N	+	+	N	N	N	N	N	N	+	N
N	+	N	N	N	N	N	N	N	+	N	N	N	N	N	N	N	+	N
N	+	+	N	N	N	N	+	N	+	N	N	N	N	N	N	N	+	N
N	N	+	+	N	+**	+**	N	+	+	+	N	+	N	N	+	N	+	N
N	N	N	N	N	N	+	N	N	N	N	N	+	N	N	N	N	N	+

coupling glucokinase activity to glucose-6-phosphate dehydrogenase (G-6-PD) activity, which in turn is coupled to the tetrazolium system (Section C, 2). Coupling may extend through more than one enzyme, as illustrated in Section C, 3.

66 / Chapter 5. Specific Electrophoretic Systems

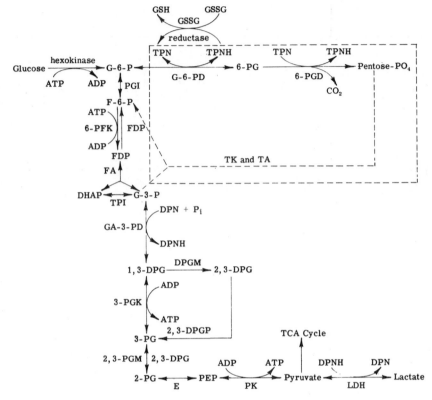

Figure 10. Metabolic scheme showing the glycolytic pathways. The pentose phosphate pathway is enclosed within the dashed lines. This figure should be used to understand the relationships among some of the enzymes described in this chapter. Abbreviations: ATP and ADP, adenosine triphosphate and diphosphate, respectively; G-6-P, glucose-6-phosphate; TPN and TPNH, oxidized and reduced triphosphopyridine nucleotide, respectively; GSSG and GSH, oxidized and reduced glutathione, respectively; G-6-PD, glucose-6-phosphate dehydrogenase; 6-PG, 6-phosphogluconate; 6-PGD, 6-phosphogluconate dehydrogenase; TK, transketolase; TA, transaldolase; F-6-P, fructose-6-phosphate; FDP, fructose-1,6-diphosphate; DHAP, dihydroxyacetone phosphate; G-3-P, glyceraldehyde-3-phosphate; DPN and DPNH; oxidized and reduced diphosphopyridine nucleotide, respectively; P_i, inorganic phosphate; 1,3-DPG, 1,3-diphosphoglycerate; 3-PG, 3-phosphoglycerate; 2-PG, 2-phosphoglycerate; 2,3-DPG, 2,3-diphosphoglycerate; PEP, phosphoenolypyruvate; PK, pyruvate kinase; LDH, lactate dehydrogenase; PGI, phosphohexose isomerase (also referred to as PHI); 6-PFK, 6-phosphofructokinase; FA, fructoaldolase; TPI, triosephosphate isomerase; GA-3-PD, glyceraldehyde-3-phosphate dehydrogenase; DPGM, diphosphoglycerate mutase; 2,3-DPGP, 2,3-diphosphoglycerate phosphatase; 3-PGK, 3 phosphoglycerate kinase; 2,3-PGM, 2,3-phosphoglycerate mutase; E, enolase; TCA, tricarboxylic acid.

As mentioned above, almost all zone electrophoresis work has employed visual methods. However, with the development of improved densitometric equipment and gel scanning devices, it should be possible to detect changes in optical density at any wavelength; this would greatly increase the versatility of the zone electrophoresis approach (see Chapter 6).

2. The Tetrazolium System

The tetrazolium dyes are of central importance in isozyme staining techniques. Any enzyme reducing either of the pyridine nucleotides (triphosphopyridine nucleotide or diphosphopyridine nucleotide) can be stained with the tetrazolium method. Further, if the enzyme under investigation (such as the glucokinase example cited above in Section C, 1) can be coupled to a pyridine nucleotide reducing enzyme, it can be stained by this method. The method always employs an electron carrier, such as phenazine methosulfate. The reduced pyridine nucleotide reduces the phenazine, which in turn reduces the tetrazolium. As the tetrazolium is reduced, it is precipitated as a formazan, which makes a bluish stain.

3. The Adenosine Triphosphate (ATP) Detecting System

As an example of the building up of systems, one on another, by coupling, the ATP detecting system may be cited (Figure 10 should be used for reference). The G-6-PD system (Section E, 1) is stained using the tetrazolium system (Section C, 2). The glucokinase system (Sections C, 1 and E, 4) is stained by coupling to the G-6-PD system. Since ATP is a required cofactor for glucokinase activity, any enzyme generating ATP, such as adenylate kinase Section E, 11) or pyruvate kinase (Section E, 21), can be coupled to the glucokinase-G-6-PD-tetrazolium system.

D. Reagents

In general, reagent grade chemicals should be used. Biochemicals in use in our laboratory for these systems are purchased from Sigma Chemical Co., St. Louis, Missouri, California Biochemical Co., Los Angeles, California, Boehringer Mannheim Co., New York, New York, K & K Laboratories, Plainview, New York, General Biochemical Co., Chagrin Falls, Ohio, and Nutritional Biochemical Co., Cleveland, Ohio. In Table III the abbreviations for some of the biochemicals, the mo-

Chapter 5. Specific Electrophoretic Systems

Table III
CHEMICALS

Chemicals	Abbreviation or chemical symbols	Molecular weight	Comments and specifications
Acetylthiocholine iodide	—	289	—
Adenosine diphosphate	ADP	449 (sodium salt)	—
Adenosine monophosphate	AMP	—	—
Adenosine triphosphate	ATP	551 (disodium salt)	—
Ammonium molybdate	$(NH_4)_6Mo_7O_{24}$	1236 (tetrahydrate)	—
Ammonium sulfide	$(NH_4)_2S$	68	—
Benzidine	—	184	—
1,5-Bis-(4-trimethylammonium-phenyl)pentan-3-one diiodide	—	—	—
Blue RR salt	—	—	Allied Chemicals
Boric acid	—	62	—
Citric acid	—	210	—
Cupric sulfate	$CuSO_4$	250 (pentahydrate)	—
2,6-Dichlorophenolindophenol	DCIP	290 (sodium salt)	—
5,5-Diethyl barbituric acid	—	184	—
Dihydroxyacetone phosphate	DHAP	184	—
3-(4,5-Dimethylthiazolyl-2)-2,5-diphenyl tetrazolium	MTT tetrazolium	414	—
1,3-Diphosphoglycerate	1,3-DPG	—	—
Diphosphopyridine nucleotide	DPN	664	—
Diphosphopyridine nucleotide (reduced)	DPNH	709	—
5,5-Dithiobis	2-Nitrobenzoic acid	396	—
Ethylenediaminetetracetate	EDTA	292 (acid)	—
Ethylenediaminetetracetate	EDTA	372 (disodium salt)	—
Fast black K	—	—	K and K Laboratories
Fructose	—	180	—
Fructose-1,6-diphosphate	F-1,6-DP	428 (tetrasodium salt)	—
Fructose-6-phosphate	F-6-P	304 (disodium salt)	Sigma Grade I, (0.9% G-6-P)
Fumaric acid	—	116	—
Glucose	—	180	—
Glucose-1,6-diphosphate	G-1,6-DP	—	—
Glucose-1-phosphate	G-1-P	304 (anhydrous)	Sigma Grade III, many lots contain enough G-1,6-DP to catalyze the PGM reaction

Table III (*Continued*)

Chemicals	Abbreviation or chemical symbols	Molecular weight	Comments and specifications
Glucose-6-phosphate	G-6-P	358 (disodium trihydrate)	—
Glucose-6-phosphate dehydrogenase	G-6-PD	—	Boehringer, 5 mg/ml, approx. 140 units/mg
L-Glutamic acid	—	169 (monosodium salt)	—
Glutathione (oxidized)	GSSG	612 (free acid)	—
Glutathione (reduced)	GSH	—	—
Glyceraldehyde-3-phosphate dehydrogenase	GA-3-PD	—	Sigma, 10 mg/ml, approx. 85 units/mg
Glyceraldehyde-3-phosphoric acid	GA-3-P	170	—
α,β-Glycerophosphate	α-GP	306 (disodium pentahydrate)	—
Glycine	—	75	—
Hexokinase	—	—	Boehringer, 10 mg/ml
Histidine hydrochloride	—	210 (monohydrate)	—
(Hydroxymethyl) aminomethane	Tris	121	—
Hypoxanthine	—	136	—
Ionagar No. 2	—	—	Colab Laboratories
DL-Isocitric acid	—	258 (trisodium salt)	—
L-Leucyl-β-naphthylamide HCl	—	293	—
Magnesium chloride	$MgCl_2$	203 (hexahydrate)	—
Maleic acid	—	116	—
DL-Malic acid	—	134	—
Malic dehydrogenase	MDH	—	Sigma, 3 mg/ml approx. 160 units/mg
Manganous chloride	$MnCl_2$	198 (tetrahydrate)	—
p-Methylaminophenol sulfate	—	344	—
α-Napthyl acetate	—	182	—
β-Napthyl acetate	—	182	—
p-Nitro blue tetrazolium	NB tetrazolium	818	—
Phenazine methosulfate	—	306	—
Phenolphthalein diphosphate	—	589 (pentasodium salt)	—

Table III (*Continued*)

Chemicals	Abbreviation or chemical symbols	Molecular weight	Comments and specifications
Phosphocreatine	—	255 (disodium salt hydrate)	—
Phospho(enol)pyruvic acid	PEP	327 (trisodium salt, pentahydrate)	—
6-Phosphogluconate	6-PG	342 (trisodium salt, anhydrous)	—
Phosphohexose isomerase	PHI	—	Sigma Grade III, crystalline
Potassium chloride	KCl	75	—
Potassium iodide	KI	166	—
Potassium phosphate	K_2HPO_4	174 (dibasic)	—
Sodium arsenate	$Na_2HA_5O_4$	312 (septahydrate)	—
Sodium barbital	—	206	—
Sodium bisulfite	$NaHSO_3$	104	—
Sodium chloride	NaCl	58	—
Sodium citrate	—	294	—
Sodium hydroxide	NaOH	40	—
Sodium phosphate	Na_2HPO_4	142 (dibasic)	—
Sodium sulfate	Na_2SO_4	142	—
Sodium thiosulfate	$Na_2S_2O_3$	248 (pentahydrate)	—
Succinic acid	—	162 (disodium salt)	—
Sulfuric acid	H_2SO_4	98	—
Tetraisopropylphosphoramide	iso-OMPA	—	K and K Laboratories
Triphosphopyridine nucleotide	TPN	765 (sodium salt)	—
Triphosphopyridine nucleotide (reduced)	TPNH	833 (tetrasodium salt)	—

lecular weights, and the specifications of some of the enzyme preparations are presented.

E. Isozyme Methods Employed in Our Laboratory

(Figure 10 will be helpful in understanding the glycolytic isozyme systems.)

1. Glucose-6-Phosphate Dehydrogenase (G-6-PD)

a. REFERENCES

This method was developed by Shows *et al.* (1964).

E. Isozyme Methods Employed in Our Laboratory / 71

b. Basis for the Staining Procedure

The site of G-6-PD isozymes are stained according to the following reactions:

$$\text{G-6-P} + \text{TPN} \xrightarrow{\text{G-6-PD}} \text{6-PG} + \text{TPNH}$$

The G-6-PD activity generates TPNH. The site of TPNH production is marked through the tetrazolium system (Section C, 2).

c. Reagents

(1) Gel buffer
 0.021 M tris
 0.02 M boric acid
 0.00068 M EDTA
The pH is adjusted to 8.6.

(2) Bridge buffer
 0.21 M tris
 0.15 M boric acid
 0.006 M EDTA
The pH is adjusted to 8.0.

(3) Staining mixture
 0.00056 M G-6-P
 0.005 M MgCl$_2$
 0.00013 M TPN
 0.00012 M NB tetrazolium
 0.00013 M phenazine methosulfate
 0.05 M tris, pH 8.0

The staining mixture should be made up shortly before it is to be used because many of the reagents are not very stable.

d. Procedure

The gel is prepared and set up in the usual way except for the addition of 10 mg of TPN per each 400 ml of molten gel before degassing. In addition, 10 mg TPN is added per each 300 ml of buffer in the cathodal bridge tray (the bridge tray with the wick connecting to the gel). The purpose of the TPN is to stabilize the G-6-PD molecule during electrophoresis and prevent its inactivation. Electrophoresis is carried out for 20 hours at approximately 10 V per linear centimeter of gel. With our electrophoretic setup, the power supply usually shows about 7 mA per gel. After completion of electrophoresis the gel is sliced and stained in the usual manner.

e. Organisms Studied

This method is generally satisfactory for mammalian erythrocytes and other tissues (Table II). An alternate method for certain organisms has been developed (see the following).

f. G-6-PD Method for Wheat, Certain Other Plants, and *Drosophila*

For work with wheat, other plants, and *Drosophila* (Table II), somewhat better results have been obtained by using an alternate method. The basic method is the same, but the amount of the various reagents and the pH differ somewhat.

(1) Gel buffer
  0.015 M tris
  0.011 M boric acid
  0.002 M EDTA
The pH is adjusted to 8.3.

(2) Bridge buffer
  0.105 M tris
  0.75 M boric acid
  0.002 M EDTA
The pH is adjusted to 8.5.

(3) Staining mixture
  0.00039 M G-6-P
  0.0048 M MgCl$_2$
  0.000052 M TPN
  0.00012 M NB tetrazolium
  0.00007 M phenazine methosulfate
  0.04 M tris
The pH is adjusted to 7.1.

The procedure, including the addition of TPN to gel and cathodal bridge buffer, is identical to that described for mammalian erythrocytes in Section E, 1, d, except that electrophoresis is usually carried out for only 17 hours.

g. Comment

Upon prolonged staining, usually a second set of bands begin to appear. In human erythrocytes these are somewhat cathodal to the G-6-PD bands and result from the enzyme 6-PGD. The G-6-PD reaction generates 6-PG which then serves as substrate for 6-PGD (see Section E, 2).

E. Isozyme Methods Employed in Our Laboratory / 73

The G-6-PD, 6-PGD, and PHI systems are identical except for the staining mixtures, and thus it is possible to study the three systems together by using multislices of the gel and staining the slices separately.

The G-6-PD system has been of major importance in biochemical genetics. As indicated in Chapter 1, the first human enzyme electrophoretic polymorphism involved this system. In addition, an inherited deficiency in the activity of this enzyme in human erythrocytes predisposes to drug-induced hemolytic anemia. The number of drugs capable of causing hemolytic anemia is quite large and includes several in common clinical use. The enzyme is controlled by an X-linked gene, not only in man, but in a number of other species. Because of its strong tendency to be X-linked, it may be speculated that the enzyme subserves some function related to sexual dimorphism. Studies of this enzyme have made valuable contributions to the understanding of X-chromosomal inactivation in mammals (Davidson et al., 1963; Brewer et al., 1967a). The G-6-PD deficiency allele is thought to reach high frequency in some populations because of a protective effect against falciparum malaria (Motulsky, 1960).

Studies by Shaw (1966) have revealed two types of G-6-PD in mouse tissues. The A type enzyme was specific for glucose-6-phosphate, while the B-type enzyme, which showed an autosomally controlled polymorphism, was equally active with glucose-6-phosphate and galactose-6-phosphate. A similar type situation in human liver was reported by Ohno et al. (1966).

A good review of the G-6-PD system and a presentation of two standardized procedures for electrophoretic study of G-6-PD may be found in a recent report of the World Health Organization (Standardization of Procedures, 1967).

2. 6-Phosphogluconate Dehydrogenase (6-PGD)

a. REFERENCES

The method is similar to that of Shows et al. (1964) for G-6-PD (Section E, 1) except that the staining reaction is carried out for 6-PGD.

b. BASIS FOR THE STAINING PROCEDURE

The sites of 6-PGD isozymes are stained according to the following reactions:

$$\text{6-PG} + \text{TPN} \xrightarrow{\text{6-PGD}} CO_2 + \text{ribose-5-P} + \text{TPNH}$$

The activity of 6-PGD generates TPNH. The site of TPNH production is marked with the tetrazolium system (Section C, 2).

c. Reagents

(1) Gel buffer
   0.021 M tris
   0.02 M boric acid
   0.00068 M EDTA
The pH is adjusted to 8.6.
(2) Bridge buffer
   0.21 M tris
   0.15 M boric acid
   0.006 M EDTA
The pH is adjusted to 8.0.
(3) Staining mixture
   0.00056 M 6-PG
   0.005 M $MgCl_2$
   0.00013 M TPN
   0.00012 M NB tetrazolium
   0.00013 M phenazine methosulfate
   0.05 M tris, pH 8.0

d. Procedure

The gel is prepared and set up in the usual way. Unlike G-6-PD, it is not necessary to add TPN for stabilization of 6-PGD, but in the case of most electrophoretic studies it apparently does no harm if it is present if the other half of the gel is to be stained for G-6-PD. Electrophoresis is carried out for 20 hours at approximately 10 V per linear centimeter of gel. With our electrophoretic setup the power supply usually shows about 7 mA per gel. After completion of electrophoresis the gel is sliced and stained in the usual manner.

e. Organisms Studied

This method is generally suitable for study of mammalian erythrocytes and other tissues (Table II). An alternate method for certain organisms has been developed (see next section).

f. 6-PGD Method for Wheat, Certain Other Plants, and *Drosophila*

Because an alternate method was developed for G-6-PD for the study of certain organisms, and because our studies of 6-PGD have been done together with G-6-PD, we have also used the alternate method for 6-PGD for the study of these same organisms (Table II).

(1) Gel buffer
   0.015 M tris
   0.011 M boric acid
   0.002 M EDTA
The pH is adjusted to 8.3.
(2) Bridge buffer
   0.105 M tris
   0.75 M boric acid
   0.002 M EDTA
The pH is adjusted to 8.5.
(3) Staining mixture
   0.00039 M 6-PG
   0.0048 M $MgCl_2$
   0.000052 M TPN
   0.00012 M NB tetrazolium
   0.00007 M phenazine methosulfae
   0.04 M tris
The pH is adjusted to 7.1.

g. Comment

The 6-PGD, G-6-PD, and PHI systems are identical except for the staining mixtures, and it is possible to study the three systems together by using multislices of the gel and staining the slices separately.

The 6-PGD system in human erythrocytes has been recently reviewed by Brewer (1969). Inherited deficiencies of the enzyme occur in human erythrocytes and an electrophoretic polymorphism exists in most human populations. In *Drosophila*, G-6-PD and 6-PGD are both sex linked, but in other organisms studied so far 6-PGD is autosomally determined.

3. Phosphohexose Isomerase (PHI)

a. References

This is an unpublished method developed in our laboratory. Another method has been developed by Detter *et al.* (1968).

b. Basis for the Staining Procedure

The site of PHI isomerase are stained according to the following reactions:

$$F\text{-}6\text{-}P \xrightarrow{PHI} G\text{-}6\text{-}P$$

$$G\text{-}6\text{-}P + TPN \xrightarrow{G\text{-}6\text{-}PD} 6\text{-}PG + TPNH$$

PHI activity is coupled to G-6-PD activity which generates TPNH. The site of TPNH production is marked through the tetrazolium system (Section C, 2, this chapter).

c. REAGENTS

(1) Gel buffer
0.021 M tris
0.02 M boric acid
0.00068 M EDTA
The pH is adjusted to 8.6.
(2) Bridge buffer
0.21 M tris
0.15 M boric acid
0.006 M EDTA
The pH is adjusted to 8.0.
(3) Staining mixture
0.00032 M F-6-P
0.005 M MgCl$_2$
0.00013 M TPN
0.00024 M MTT tetrazolium
0.00013 M phenazine methosulfate
0.05 mg G-6-PD
0.05 M tris, pH 8.0

The staining mixture should be made up shortly before use since many of the reagents are not very stable.

d. PROCEDURE

The gel is prepared and set up in the usual way. This method can be conveniently combined with the G-6-PD method (Section E, 1, this chapter), but unlike G-6-PD, and like the 6-PGD method (Section E, 2, this chapter), it does not require TPN in the gel and bridge buffers. The addition of TPN however does not harm the PHI patterns which have been studied. Electrophoresis is carried out for 20 hours at approximately 10 V per linear centimeter of gel. With our electrophoretic setup the power supply usually shows about 7 mA per gel. After completion of electrophoresis the gel is sliced and stained in the usual manner.

e. ORGANISMS STUDIED

This method is satisfactory for human erythrocytes, wheat seeds, and *Drosophila* (Table II). It has not been tried with other organisms.

E. Isozyme Methods Employed in Our Laboratory

f. Comment

The PHI, G-6-PD, and 6-PGD systems are identical except for the staining mixture, and thus it is possible to study the three systems together by using multislices of the gel and staining the slices separately.

A deficiency of PHI in human erythrocytes has been associated with hemolytic anemia (Baughan *et al.*, 1968). Detter *et al.* (1968) have studied this same family with a different electrophoretic method and have found the patient to be heterozygous for two different rare alleles, each associated with reduced PHI activity.

4. Glucokinase (Hexokinase)

a. References

This basic method was developed by Eaton *et al.* (1966), and modified by Brewer and Knutsen (1968). The method described here is from the latter publication.

b. Basis for Staining Procedure

The sites of glucokinase isozymes are stained according to the following reactions:

$$\text{Glucose} + \text{ATP} \xrightarrow{\text{glucokinase}} \text{G-6-P} + \text{ADP}$$

$$\text{G-6-P} + \text{TPN} \xrightarrow{\text{G-6-PD}} \text{6-PG} + \text{TPNH}$$

Glucokinase activity is coupled to G-6-PD activity which generates TPNH. The site of TPNH production is marked through the tetrazolium system (Section C, 2).

c. Reagents

(1) Gel buffer
  0.021 M tris
  0.02 M boric acid
  0.001 M EDTA
The pH is adjusted to 8.4.
(2) Bridge buffer
  0.21 M tris
  0.15 M boric acid
  0.006 M EDTA
The pH is adjusted to 8.0.
(3) Staining mixture
  0.0083 M glucose

0.01 M $MgCl_2$
0.0024 M ATP
0.00013 M TPN
0.00024 M MTT tetrazolium
0.00024 M phenazine methosulfate
0.05 mg G-6-PD/100 ml staining solution
0.05 M tris, pH 8.4

The staining mixture should be made up shortly before use since many of the reagents are not very stable.

d. PROCEDURE

The gel is prepared and set up in the usual way. Electrophoresis is carried out for 18–20 hours at approximately 12 V per linear centimeter of gel. With our electrophoretic setup the power supply usually shows about 13 mA per gel. After completion of electrophoresis the gel is sliced and stained in the usual manner.

e. ORGANISMS STUDIED

This method is generally satisfactory for mammalian erythrocytes and tissues, wheat seeds and leaves, and for many plants (Table II). An alternate method is used for *Drosophila* (Section E, 4, f, below) which uses a higher concentration of glucose in the staining mixture.

f. GLUCOKINASE METHOD FOR *Drosophila*

The gel and bridge buffers and the electrophoretic procedure are the same as above. We have obtained better results with *Drosophila*, however, by employing a tenfold higher concentration of glucose (0.083 M) and a pH of 8.0 in the staining mixture.

g. COMMENT

The glucokinase and fructokinase systems are identical except for the staining mixtures and it is possible to study the two systems together by using multislices of the gel and staining the slices separately. We have not attempted to study the fructokinase isozymes of organisms other than *Drosophila*, but the method should work satisfactorily for other organisms.

The glucokinase isozymes of red blood cells of the human are not very active. Therefore, it take 2 or 3 hours of staining at 37°C to bring out the full complement of bands. Holmes *et al.* (1967), using a different method with considerably less resolving power, reported an apparent association between a hexokinase isozyme band and fetal hemoglobin.

E. Isozyme Methods Employed in Our Laboratory / 79

This claim appears, however, to be incorrect (Brewer and Knutsen, 1968; Kaplan and Beutler, 1968). Wheat leaves and wheat seeds also do not have strong glucokinase bands. Many species of *Drosophila*, on the other hand, have very prominent glucokinase isozyme bands. Current studies indicate that there may be at least 4 genetic loci, 2 of which are segregating, controlling glucokinase isozymes in *D. robusta* (Knutsen *et al.*, 1969).

Glucokinase may be a useful isozyme system with which to study leukemia. Differences among types of leukemia have been noted in the glucokinase isozyme patterns (Eaton *et al.*, 1966).

5. Fructokinase

a. References

This method was developed by Knutsen *et al.* (1969).

b. Basis for the Staining Procedure

The sites of fructokinase isozymes are stained according to the following reactions:

$$\text{Fructose} \xrightarrow{\text{fructokinase}} \text{F-6-P} + \text{ADP}$$

$$\text{ATP} + \text{F-6-P} \xrightarrow{\text{phosphohexose isomerase}} \text{G-6-P}$$

$$\text{G-6-P} + \text{TPN} \xrightarrow{\text{G-6-PD}} \text{6-PG} + \text{TPNH}$$

Fructokinase activity is coupled to phosphohexose isomerase and G-6-PD, which generates TPNH. The site of TPNH production is marked through the tetrazolium system (Section C, 2).

c. Reagents

(1) Gel buffer
  0.021 M tris
  0.02 M boric acid
  0.001 M EDTA
The pH is adjusted to 8.4.
(2) Bridge buffer
  0.21 M tris
  0.15 M boric acid
  0.006 M EDTA
The pH is adjusted to 8.0.
(3) Staining mixture (for *Drosophila*)
  0.083 M fructose

0.01 M MgCl$_2$
0.0024 M ATP
0.00013 M TPN
0.00024 M MTT tetrazolium
0.00024 M phenazine methosulfate
0.05 mg G-6-PD/100 ml staining solution
0.11 mg phosphohexose isomerase/100 ml staining solution
0.05 M tris, pH 8.0

The staining mixture should be made up shortly before use since many of the reagents are not very stable.

d. Procedure

The gel is prepared and set up in the usual way. Electrophoresis is carried out for 18–20 hours at approximately 12 V per linear centimeter of gel. With our electrophoretic setup, the power supply usually shows about 13 mA per gel. After completion of electrophoresis the gel is sliced and stained in the usual manner.

e. Organisms Studied

This method has been applied only to *Drosophilia*.

f. Comment

The glucokinase and fructokinase systems are identical except for the staining reaction and hence it is possible to study the two systems together by using multislices of the gel and staining the slices separately. We have employed this dual staining method in the study of *Drosophila*. Note that the staining mixture specified in this method is identical to the glucokinase staining method for *Drosophila* (Section E, 4, f) except for the substrate, fructose, and the presence of phosphohexose isomerase in the solution. In the case of *Drosophila* certain isozymes have dual specificity, while others are primarily only glucokinases. We have not found any *Drosophila* isozymes which have only fructokinase activity. Similar dual studies of the hexokinase isozymes of other organisms should be fairly straightforward with these techniques.

6. Phosphoglucomutase (PGM)

a. References

This is the method of Spencer *et al.* (1964), except that vertical rather than horizontal electrophoresis is employed.

b. Basis for the Staining Procedure

The sites of PGM isozymes are stained according to the following reactions:

$$\text{G-1-P} \xrightarrow[\text{G-1,6-DP}]{\text{PGM}} \text{G-6-P}$$

$$\text{G-6-P} + \text{TPN} \xrightarrow{\text{G-6-PD}} \text{6-PG} + \text{TPNH}$$

A cofactor, G-1,6-DP, is required for PGM activity. This cofactor is often present as a contaminant in G-1-P preparations. PGM activity is coupled to G-6-PD activity which generates TPNH. The site of TPNH production is marked through the tetrazolium system.

c. Reagents

(1) Gel buffer
   0.01 M tris
   0.01 M maleic acid
   0.001 M EDTA
   0.001 M MgCl$_2$

The pH is adjusted to 7.6 with 4 N NaOH. Some of the reagents (notably EDTA) will not go into solution until the NaOH is added.

(2) Bridge buffer
   0.1 M tris
   0.1 M maleic acid
   0.01 M EDTA
   0.01 M MgCl$_2$

The pH is adjusted to 7.6 with 4 N NaOH. Some of the reagents (notably EDTA) will not go into solution until NaOH is added. Note that the gel buffer is a 1:10 dilution of the bridge buffer.

(3) Staining mixture
   0.0046 M glucose-1-phosphate (Grade III from Sigma Chemical Co. usually contains enough G-1,6-DP in the preparation to catalyze the reaction; see Section E, 6, b)
   0.01 M MgCl$_2$
   0.0012 M TPN
   0.00033 M phenazine methosulfate
   0.00024 M MTT tetrazolium
   0.075 mg G-6-PD/100 ml staining solution
   0.03 M tris pH 8.0

d. Procedure

The gel is prepared and set up in the usual way. Electrophoresis is carried out for 18–20 hours at 8–10 V per linear centimeter of gel, at approximately 15 mA per gel. After completion of electrophoresis the gel is sliced and stained in the usual manner.

e. ORGANISMS STUDIED

This method is very good for mammalian erythrocytes. We have applied it as well to wheat seeds and leaves, to many other plants, and to *Drosophila* (see Table II).

f. COMMENT

A cofactor, G-1,6-DP is required for PGM activity. This cofactor is not commercially available. However, most preparations of G-1-P, if they are not too highly purified, have sufficient G-1,6-DP as a contaminant to allow the reaction to proceed.

Mammalian erythrocytes have at least two autosomal loci governing PGM isozymes, one of which (PGM_1) is polymorphic (Spencer *et al.*, 1964) in a number of human populations. The PGM_2 locus shows occasional variants (Hopkinson and Harris, 1965; Brewer *et al.*, 1967b). The two loci do not seem to be closely linked.

It should be noted that because of identical gel and bridge buffers the study of PGM isozymes can be conveniently combined with the study of achromatic regions, LDH, or ATPase.

7. Achromatic Regions (Indophenol Oxidase Activity)

a. REFERENCES

This method was developed by Brewer (1967).

b. BASIS FOR THE STAINING PROCEDURE

A starch gel incubated with phenazine methosulfate and a tetrazolium dye will gradually develop a bluish background if exposed to light. This bluish background can be noted in any of the tetrazolium staining methods reported in this chapter and seems to result from an ultraviolet catalyzed reduction of the tetrazolium dye. This reaction is accelerated in alkaline solution. Erythrocytes and many other mammalian tissues, as well as *Drosophila* and many plants, possess enzymes which are capable of either preventing this reaction or reversing it. These protein bands show up as light areas in the bluish background of the gel. These light, or achromatic areas, are the indicators of the presence of the enzyme.

c. Reagents

(1) Gel buffer
0.01 M tris
0.01 M maleic acid
0.001 M EDTA
0.001 M MgCl$_2$

Adjust the pH to 7.6 with 4 N sodium hydroxide. Note that some of these reagents will not go into solution until the sodium hydroxide is added

(2) Bridge buffer
0.1 M tris
0.1 M maleic acid
0.01 M EDTA
0.01 M MgCl$_2$

Adjust the pH to 7.6 with 4 N sodium hydroxide. Note that the gel buffer is a 1:10 dilution of the bridge buffer.

(3) Staining mixture
0.01 M MgCl$_2$
0.00033 M phenazine methosulfate
0.00024 M MTT tetrazolium
0.03 M tris, pH 8.0

The staining mixture should be made up shortly before use since certain of the reagents are not very stable.

d. Procedure

The gel is prepared and set up in the usual way. Electrophoresis is carried out for 18–20 hours at 8–10 V per linear centimeter of gel at approximately 15 mA per gel. After completion of electrophoresis the gel is sliced and stained in the usual manner, except that it is of importance to expose the gel to light, so that the gel developes a bluish background.

e. Organisms Studied

We have applied this method primarily to the study of human erythrocytes and tissues (Brewer, 1967). However, the achromatic regions have been noted in most tissues of all organisms studied. It is also seen in plants (see Table II).

f. Comment

Operationally, the enzyme responsible for one set of achromatic bands of human erythrocytes has been shown to be an indophenol oxidase (Brewer, 1967). Enzymes of this general class are considered soluble oxidases capable of catalyzing aerobic oxidation of various intracellular substances, including reduced cytochrome c (Yonetani, 1963). An autosomally determined electrophoretic variant of the enzyme has been found in human erythrocytes (Brewer, 1967).

It should be noted that because of identical gel and bridge buffers the study of achromatic regions can be conveniently combined with the study of PGM, LDH, and ATPase isozymes by using multislices of the gel and staining the slices separately.

8. Lactate Dehydrogenase (LDH)

a. REFERENCES

The use of this buffer system with LDH was developed in our laboratory. Many other methods have been reported (reviews in Latner and Skillen, 1968; Wilkinson, 1965).

b. BASIS FOR THE STAINING PROCEDURE

The sites of LDH isozymes are stained according to the following reaction:

$$\text{Lactate} + \text{DPN} \xrightarrow{\text{LDH}} \text{pyruvate} + \text{DPNH}$$

Lactate dehydrogenase activity generates DPNH. The site of DPNH production is marked through the tetrazolium system (Section C, 2).

c. REAGENTS

(1) Gel buffer
  0.01 M tris
  0.01 M maleic acid
  0.001 M EDTA
  0.001 M $MgCl_2$

The pH is adjusted to 7.6 with 4 N sodium hydroxide. Note that some of these reagents will not go into solution until the sodium hydroxide is added.

(2) Bridge buffer
  0.1 M tris
  0.1 M maleic acid
  0.01 M EDTA
  0.01 M $MgCl_2$

The pH is adjusted to 7.6 with 4 N sodium hydroxide. Note that the gel buffer is a 1:10 dilution of the bridge buffer.

(3) Staining mixture
0.535 M sodium lactate
0.000376 M DPN
0.000163 M phenazine methosulfate
0.00031 M NB tetrazolium
0.025 M tris, pH 7.5

The staining mixture should be made up shortly before use since some of the reagents are not very stable.

d. Procedure

The gel is prepared and set up in the usual way. Electrophoresis is carried out for 18–20 hours with approximately 8–10 V per linear centimeter of gel at about 15 mA per gel. After completion of electrophoresis the gel is sliced and stained in the usual manner.

e. Organisms Studied

Our studies have been primarily of erythrocytes and certain plants (Table II) but LDH has been studied in a variety of organisms in other laboratories.

f. Comment

The LDH, PGM, ATPase, and achromatic regions systems are identical except for the staining mixtures; it is possible to study the four systems together by using multislices of the gel and staining the slices separately.

The LDH system has traditionally been one of the cornerstones of isozymology (Chapter 1). It has been the system which has been most useful clinically (Chapter 6). The molecule consists of a tetramer of two types of randomly combining subunits. This generates five isozymes in tissues where both subunits are synthesized. Some tissues synthesize more of one type of subunit than the other, leading to a predominance of certain isozymes. Tissue damage, with release of the cellular enzymes into the serum, leads to the appearance of the isozyme pattern of the particular damaged tissue. In myocardial infarction, LDH-1 predominates in the serum; in hepatocellular damage, LDH-5 predominates (Chapter 6).

Interesting metabolic differences have been observed between the LDH isozymes. LDH-1 is strongly inhibited by low concentrations of pyruvate, preventing accumulation of pyruvate if a tissue has predominantly this isozyme, for example, heart muscle. In such a case, complete oxidation via the tricarboxylic acid cycle is to be expected. LDH-5, on the other hand, is not inhibited by low concentrations of pyruvate. Tissues with predominantly LDH-5, such as skeletal muscle, can develop an oxygen debt under anaerobic conditions by accumulation of lactate (Cahn et al., 1962). Studies such as this begin to place the role of isozymes in proper physiological perspective.

Another important use of the LDH system has been illustrated in the study of ontogeny, first initiated by Markert and Moller (1959). Many additional studies of LDH and other isozyme systems have shown very frequent occurrence of ontogenetic variation in isozyme patterns.

The LDH system was also important in the early demonstration of the use of genetic variation to understand enzyme structure. It was shown that a mutation (affecting 1 subunit) produced 10 new isozymes, the structure of which could be explained by a tetramer-2 subunit hypothesis (Shaw and Barto, 1963). This topic is discussed in more detail in Chapter 6.

9. Esterase

a. REFERENCES

This is basically the method of Tashian and Shaw (1962) and Tashian (1969). The staining mixture is essentially the same as that described by Markert and Hunter (1959).

b. BASIS FOR THE STAINING PROCEDURE

The sites of esterase isozymes are stained by the coupling of α-napthol with a diazonium salt (Blue RR) after the α-napthol is liberated from α-napthyl acetate by the esterase activity.

c. REAGENTS

(1) Gel buffer
  0.02 M boric acid
The pH is adjusted to 8.62 with 2 N sodium hydroxide.
(2) Bridge buffer
  0.3 M boric acid
  0.03 M sodium chloride
The pH is adjusted to 8.0 with 2 N sodium hydroxide.
(3) Staining mixture
  0.006 M α-napthyl acetate (it is convenient to make up 8 ml of 1% α-napthyl acetate for every 100 ml of staining solution; since the α-napthyl acetate is not very soluble it should be made up in 50% water and 50% acetone; the α-napthyl acetate can be dissolved in acetone first, and then the water can be added)
  200 mg Blue RR salt/100 ml staining solution
  0.08 M tris, pH 7.0

d. Procedure

For best results use two anodal trays and only one cathodal tray. The electrode is placed in the cathodal tray at the top containing the boric acid and sodium chloride mixture. Of the anodal trays, the bridge buffer tray (in which the gel sets) contains the boric acid and sodium chloride mixture, while the electrode tray contains 10% sodium chloride.

Electrophoresis is carried out for about 17 hours at approximately 8 V per linear centimeter of gel with a current of about 10 mA per gel. After completion of electrophoresis the gel may be sliced in the usual manner.

For staining, mix and pour the staining mixture (filtered through glass wool) over the gel and let stand at room temperature (25°–37°C). The staining solution should be changed as it becomes dark or cloudy (approximately every 30 minutes). The gel can then be placed in 50% ethanol. After 24 hours of exposure to ethanol the gels may be kept for several days in water, or wrapped in plastic wrapping paper and kept indefinitely at refrigerator temperatures.

e. Organisms Studied

This method has wide applicability to blood, animal tissues, wheat and other plants, and *Drosophila* (Table II).

f. Comment

The esterase and CA systems are very similar except for the staining mixtures, and it is possible to study the two systems together by using multislices of the gel and staining the slices separately.

The esterase system is a nonspecific method, in the sense that the ester substrate used for identifying the isozymes may have little to do with the biological substrate of the enzymes, and the stain may reveal several classes and types of enzymes bearing varying or no relation to one another. It is very common for the products of several genetic loci to be demonstrated in such esterase zymograms. Other esterase substrates, as well as inhibitors, can be used in the differentiation and study of this group of enzymes. [For reviews see Tashian (1969) and Latner and Skillen (1968).]

In general, three groups of esterases are present in biological systems. These are called the arylesterases or A (aromatic) esterases, the aliesterases or B esterases, and the cholinesterases. The A esterases hydrolyze aromatic substrates more readily than aliphatic substrates, and vice versa, although there is cross reactivity. Cholinesterases also show activity toward these substrates, although they are most active against choline esters. Thus, with an aromatic substrate such as used here, all three groups of enzymes may stain. A specific method for cholinesterase is presented later in the chapter (Section E, 2, 9).

The historical significance of the esterases in isozymology has been discussed in Chapter 1. Of particular importance in the field of pharmacogenetics was the discovery of a genetic polymorphism involving human serum cholinesterase. When the muscle relaxant, suxamethonium, was used in surgery, a few individuals developed prolonged apnea. Such persons have been found to have low activity of serum cholinesterase (Bourne *et al.*, 1952; Evans *et al.*, 1952). The serum cholinesterase of such individuals is unusually sensitive to dibucaine (Kalow and Genest, 1957).

10. Carbonic Anhydrase (CA)

a. REFERENCES

This is essentially the method of Tashian and Shaw (1962) and Tashian (1969).

b. BASIS FOR THE STAINING PROCEDURE

The sites of CA isozymes are stained by taking advantage of the esterase activity of the enzyme. CA has a greater activity toward β-napthyl acetate than α-napthyl acetate, so the former compound is used. β-napthol is liberated by the CA, and complexes with Blue RR salt. It is convenient, at least in the case of erythrocytes, to stain for both esterase and CA activity simultaneously. This is done by including both α-napthyl acetate and β-napthyl acetate in the staining mixture. A pinkish-purple color results from the β-napthol/Blue RR complex while the other esterases will stain bluish-gray as a result of the α-napthol/Blue RR complex. A definitive test for the identification of CA activity is to include 10^{-4} M acetazolamide (Diamox) in the staining mixture and compare with a gel stained in the usual manner. Acetazolamide is a specific inhibitor of CA.

An alternative method consists of preparing a 0.1% bromthymol blue solution in 0.1 M sodium barbital buffer, pH 9.0. A piece of filter paper is dipped in this solution and placed over the gel. After several minutes, the paper is removed and the gel is bathed with carbon dioxide from a cylinder of compressed carbon dioxide. The CA isozymes appear as yellowish green bands as a result of the action of CA to form carbonic acid.

The basic reaction of CA is as follows:

$$H_2O + CO_2 \leftrightarrow H_2CO_3$$

The esterase activity of CA is analogous to the backward reaction.

c. REAGENTS

(1) Gel buffer
0.02 M boric acid
The pH is adjusted to 8.76 with 2 N sodium hydroxide.
(2) Bridge buffer
0.3 M boric acid
0.03 M sodium chloride
The pH is adjusted to 8.0 with 2 N sodium hydroxide.
(3) Staining mixture
0.006 M β-napthyl acetate (instructions in Section E, 9, c, (3), also apply to the preparation of this solution)
200 mg Blue RR salt
0.08 M tris, pH 7.0

d. PROCEDURE

The procedure is similar to that in Section E, 9, d for esterases. Staining for CA result in a transitory band and thus the band should be recorded or photographed as it develops.

e. ORGANISMS STUDIED

This method has been applied primarily to mammalian erythrocytes (Table II).

f. COMMENT

The carbonic anhydrase and esterase (Section E, 9) methods can be conveniently combined for study in the same electrophoretic run. This can be accomplished by using two substrates for staining (see Section E, 10, b) or by staining two slices of the gel individually. Diamox can be used as an inhibitor of CA to test for its activity, and the bromthymol blue–CO_2 procedure described in Section E, 10, b is specific for CA. For combined study the gel buffer of CA (pH 8.76) should be used for best results, at least in erythrocytes.

The historical significance of CA has been discussed in Chapter 1. It is the first red cell esterase to have a physiological role shown, and also the first enzyme of diploid organisms in which an amino acid subsituation has been shown (Tashian et al., 1966).

Almost without exception, primate erythrocytes have two CA isozymes referred to as CA I and CA II (Tashian, 1969). These are controlled by two distinct unlinked autosomal loci. CA I and CA II have somewhat different kinetic characteristics. These isozymes always occur in the same relationship to one another on the gel (CA I always anodal to CA II) across a large number of primate species. This suggests the possibility that the charge on the respective molecules may have something to do with disparate functions.

11. Adenylate Kinase

a. REFERENCE

This is the method of Fildes and Harris (1966) except that vertical rather than horizontal electrophoresis is employed.

b. BASIS FOR THE STAINING PROCEDURE

The sites of AK isozymes are stained according to the following reactions:

$$2ADP \xrightarrow{AK} ATP + AMP$$

$$ATP + glucose \xrightarrow{hexokinase} ADP + G\text{-}6\text{-}P$$

$$G\text{-}6\text{-}P + TPN \xrightarrow{G\text{-}6\text{-}PD} 6\text{-}PG + TPNH$$

AK activity is coupled to hexokinase and G-6-PD, which generate TPNH. The site of TPNH production is marked through the tetrazolium system (Section C, 2).

c. REAGENTS

(1) Gel buffer
  0.005 M histidine
The pH is adjusted to 7.0 with 2 N sodium hydroxide.
(2) Bridge buffer
  0.41 M sodium citrate
The pH is adjusted to 7.0 with 0.41 M citric acid.
(3) Staining mixture
  0.001135 M ADP
  0.01 M glucose
  0.011 M MgCl$_2$
  0.00043 M TPN
  0.000392 M phenazine methosulfate
  0.00029 M MTT tetrazolium
  0.05 mg G-6-PD
  0.1 mg hexokinase
  0.75 gm Ionagar/100 ml of staining solution
  0.1 M tris, pH 8.0

Because of the double coupling to detect the bands with soluble intermediates, an agar overlay method is used. The Ionagar is dissolved in ½–¾ of the tris buffer and heated to boiling. This mixture is then allowed to cool to 45°C. While it is cooling, the other reagents of the staining mixture are added to the remaining tris buffer and this mixture is stirred thoroughly until all constituents are dissolved. The two mixtures are then added together and stirred. This solution is now ready to pour as an agar overlay.

d. Procedure

The gel is prepared and set up in the usual way. Electrophoresis is carried out for 4 hours at approximately 8–10 V per linear centimeter of gel and about 15–20 mA per gel. After completion of electrophoresis, the gel is sliced in the usual manner and the staining mixture containing agar prepared as described in the preceding section. The liquid staining mixture (temperature 40°–45°C) is poured over the sliced gel in a staining tray. As the agar solution cools to room temperature it will solidify. The gels may then be incubated in an incubator at 37°C for 2 hours, or until staining is complete.

e. Organisms Studied

This method is generally satisfactory for mammalian erythrocytes, *Drosophila*, wheat seeds and leaves, and for many other plants (see Table II).

f. Comment

Ordinarily an agar overlay method is not required if the colored product formed is insoluble, such as the formazan precipitate of the tetrazolium stain. However, because the coupling reaction involves three enzymes with soluble reactants in each case, the agar overlay helps fix these products in sufficient concentration to allow color to develop.

It should be noted that the study of AK can be conveniently combined with the study of a great number of other enzymes which use the same gel and bridge buffers. These include alkaline phosphatase, α-GPD, GA-3-PD, ICD, SDH, LAP, peroxidase, fumarase, aldolase, PK, and CK.

Human erythrocytes show an autosomally determined genetic polymorphism of AK involving the Caucasian, but not the Negro race (Fildes and Harris, 1966; Bowman et al., 1967; Brewer et al., 1967b).

It is of biochemical and possible evolutionary interest that in many species, multiple kinase activities seem to reside in identical molecular species. We can be reasonably certain that in mammalian erythrocytes, heart and liver, and in wheat seeds (Table II), both AK activity and PK (Section E, 21) activity are shared in one or more isozymes. The isozyme patterns in these tissues are identical and both activities are known to be present in these tissues. Of all the species tested only *Drosophila* had a specific PK band that did not also show AK activity. A similar story may unfold for CK (Section E, 22). CK isozyme patterns are identical for AK patterns across all species and organisms tested except skeletal muscle (Table I). However, in this case we cannot be certain of the presence of CK activity in all of the tissues studied, and we may, in some cases, simply be picking up AK isozymes only because of the necessary presence of ADP in the CK staining mixture.

12. Alkaline Phosphatase

a. References

The staining method is a modification of the method of Boyer (1961). Application of the buffer system to this enzyme system was developed in our laboratory.

b. Basis for the Staining Procedure

The phosphatase activity frees napthol from α-napthyl phosphate. The liberated napthol is stained with Blue RR. To fall within the definition of alkaline phosphatases, the staining reaction should be carried out at alkaline pH.

c. Reagents

(1) Gel buffer
  0.005 M histidine
The pH is adjusted to 7.0 with 2 N sodium hydroxide.
(2) Bridge buffer
  0.41 M sodium citrate
The pH is adjusted to 7.0 with 0.41 M citric acid.
  (3) Staining mixture
    50 mg sodium α-napthyl phosphate/100 ml staining solution
    500 mg polyvinylpyrolidone/100 ml of staining solution
    0.0005 M MgCl$_2$
    50 mg Blue RR salt/100 ml of staining solution
    0.3 M NaCl
    0.01 M tris, pH 8.5
The sodium chloride and magnesium chloride are dissolved in the tris buffer before adding the other reagents.

d. Procedure

The gel is prepared and set up in the usual way. Electrophoresis is carried out for 4 hours at approximately 8–10 V per linear centimeter of gel and about 15–20 mA per gel. After completion of electrophoresis the gel is sliced and stained in the usual manner. At the end of incubation the staining mixture may be removed and replaced with 50% ethanol. After 24 hours of exposure to ethanol the gels may be kept for several days in water, or wrapped in plastic wrapping paper and kept indefinitely at refrigerator temperatures.

e. Organisms Studied

We have used this method for *Drosophila*, wheat seeds and leaves, and for many other plants (see Table II).

f. Comment

It may be noted that because of identical buffer systems, the study of alkaline phosphatase isozymes can be conveniently combined with the study of AK, α-GPD, GA-3-PD, ICD, SDH, LAP, peroxidase, fumarase, aldolase, PK, and CK.

The alkaline phosphatase system is a nonspecific method, in the sense that the phosphate substrate used for identifying the isozymes may have little to do with the biological substrates of the enzymes, and the stain may reveal several classes and types of enzymes bearing varying or no relation to one another. The phosphatases tend to show broad substrate specificity; many other substrates besides the one employed here have been used for the determination of alkaline phosphatase activity. Included within this list would be relatively specific phosphate compounds which may reveal somewhat more specific phosphatases. Examples are given in the next paragraph. The use of multiple substrates, Km's, inhibitors, or other kinetic characteristics have not been of a great deal of help in delineating dIifferent serum alkaline phosphatase isozymes. Much of this work has been done with the objective of improving the methods for identifying tissues of origin of different serum phosphatases. The clinical usefulness of serum alkaline phosphatase isozyme patterns has been reviewed by Latner and Skillen (1968).

Certain alkaline phosphatases have been identified with more specific substrates, including substrates which may be related to the physiological functions of the enzymes. Usually, however, these phosphatases also show activity toward a number of phosphate compounds. Examples are nucleoside diphosphatases and thiamine pyrophosphatases (Allen, 1963;

Allen and Hynick, 1963), and six different glycerophosphatases in rat liver (Sandler and Bourne, 1961, 1962).

Some evidence of genetic variation in placental alkaline phosphatases has been obtained. During the last 6 weeks of pregnancy a serum alkaline phosphatase isozyme appears which shows variation between women (Boyer, 1961). The same type pattern is found in the placenta (Latner, 1965), which is presumably the source of the serum enzyme.

Variation in leukocyte alkaline phosphatase isozyme patterns in leukemia have been reported (Robinson et al., 1965).

Alkaline phosphatases may be of considerable importance in plants, because of variation in soil phosphates and the critical need by plants for phosphate. In wheat, an almost obligatory self-pollinating organism, alkaline phosphatase was the only one of twelve isozyme systems studied which showed variation between the diploid genomes imbedded in a hexaploid plant (Brewer et al., 1969); this suggests that variations in types of alkaline phosphatases may be of selective value. (See Chapter 6 for fuller discussion.)

13. α-Glycerophosphate Dehydrogenase (α-GPD)

a. REFERENCE

This is an unpublished method developed in our laboratory.

b. BASIS FOR THE STAINING PROCEDURE

The sites of α-GPD isozymes are stained according to the following reaction:

$$\alpha\text{-GP} + \text{DPN} \xrightarrow{\alpha\text{-GPD}} \text{dihydroxyacetone PO}_4 + \text{DPNH}$$

α-GPD activity generates DPNH. The site of DPNH production is marked through the tetrazolium system (Section C, 2).

c. REAGENTS

(1) Gel buffer
 0.005 M histidine
The pH is adjusted to 7.0 with 2 N sodium hydroxide.
(2) Bridge buffer
 0.41 M sodium citrate
The pH is adjusted to 7.0 with 0.41 M citric acid.
(3) Staining mixture
 0.1 M pentahydrate disodium α,β-glycerophosphate

0.001 M DPN
0.000163 M phenazine methosulfate
0.00043 M NB tetrazolium
0.05 M tris, pH 7.0

The tris and α,β-glycerophosphate are mixed, and the pH is readjusted to 7.0. The other reagents are added to this mixture.

d. Procedure

The gel is prepared and set up in the usual way. Electrophoresis is carried out for 4 hours at approximately 8–10 V per linear centimeter of gel at about 15–20 mA per gel. After completion of electrophoresis the gel is sliced and stained in the usual manner. The gels may be incubated at 37°C for approximately 3 hours, or until staining is complete. At the end of incubation the staining mixture may be removed and replaced with 50% ethanol. After 24 hours of exposure to ethanol the gels may be kept for several days in water, or wrapped in plastic wrapping paper and kept indefinitely at refrigerator temperatures.

e. Organisms Studied

We have used this method for mammalian liver and heart, *Drosophila*, wheat seeds, and wheat leaves (see Table II).

f. Comment

The study of α-GPD isozymes may be conveniently combined with other methods which use the same buffer system. These are AK, alkaline phosphatase, GA-3-PD, ICD, SDH, LAP, peroxidase, fumarase, aldolase, PK, and CK.

14. Glyceraldehyde-3-Phosphate Dehydrogenase (GA-3-PD)

a. References

The application of this buffer system to GA-3-PD has been developed in our laboratory. It has been published in part (Brewer and Sing, 1969). For additional methodology see Williams (1964).

b. Basis for the Staining Procedure

The sites of GA-3-PD isozymes are stained according to the following reaction:

$$\text{GA-3-P} + \text{DPN} \xrightarrow{\text{GA-3-PD}} \text{1,3-DPG} + \text{DPNH}$$

GA-3-PD activity generates DPNH. The sites of DPNH production are marked through the tetrazolium system (Section C, 2).

c. Reagents

(1) Gel buffer
0.005 M histidine
The pH is adjusted to 7.0 with 2 N sodium hydroxide.
(2) Bridge buffer
0.41 M sodium citrate
The pH is adjusted to 7.0 with 0.41 M citric acid.
(3) Staining mixture
0.1 ml glyceraldehyde-3-phosphoric acid/100 ml of staining solution
0.0003 M DPN
0.0048 M sodium arsenate
0.000163 M phenazine methosulfate
0.00048 M MTT tetrazolium
0.1 M tris, pH 8.5

An agar overlay method (such as described for AK, Section E, 11) has also been used with this method to produce less diffuse and stronger bands.

d. Procedure

The gel is prepared and set up in the usual way. Electrophoresis is carried out for 4 hours at approximately 8–10 V per linear centimeter of gel and about 15–20 mA per gel. After completion of electrophoresis the gel is sliced in the usual manner and stained in the usual manner or under an agar overlay for about 3 hours or until staining is complete.

e. Organisms Studied

We have used this method for mammalian erythrocytes, wheat seeds, leaves, and for many other plants (see Table II).

f. Comment

It should be noted that the study of GA-3-PD can be conveniently combined with the study of other enzymes described in this chapter employing the same buffer system. These include AK, alkaline phosphatase, α-GPD, ICD, SDH, LAP, peroxidase, fumarase, aldolase, PK, and CK.

Isozymes of GA-3-PD have been studied in group D *Streptococci* (Williams, 1964) and in turtle, perch, trout, spinach, and yeast by H. G. Lebhery (personal communication).

15. Isocitrate Dehydrogenase (ICD)

a. REFERENCES

The application of this buffer system to ICD was developed in our laboratory (Brewer and Sing, 1969). For additional methods and references see Tsao (1960), Bell and Baron (1962), Campbell and Maas (1962), and Henderson (1965).

b. BASIS FOR THE STAINING PROCEDURE

The sites of ICD isozymes are stained according to the following reaction:

$$\text{Isocitrate} + \text{TPN} \xrightarrow{\text{ICD}} \text{oxalosuccinate} + \text{TPNH}$$

ICD activity generates TPNH. The sites of TPNH production is marked through the tetrazolium system (Section C, 2). DPN may also be used instead of TPN to contrast the coenzyme specificity (see Section E, 15, f).

c. REAGENTS

(1) Gel buffer
    0.005 M histidine
The pH is adjusted to 7.0 with 2 N sodium hydroxide.
(2) Bridge buffer
    0.41 M sodium citrate
The pH is adjusted to 7.0 with 0.41 M citric acid.
(3) Staining mixture
    0.0077 M isocitric acid trisodium salt
    0.00026 M TPN
    0.001 M MnCl$_2$
    0.00046 M phenazine methosulfate
    0.00021 M NB tetrazolium
    0.2 M tris, pH 8.0

d. PROCEDURE

The gel is prepared and set up in the usual way. Electrophoresis is carried out for 4 hours at approximately 8–10 V per linear centimeter of gel at about 15–20 mA per gel. After completion of electrophoresis the gel is sliced and stained in the usual manner. At the end of incubation the staining mixture may be removed and replaced with 50% ethanol. After 24 hours of exposure to ethanol the gels may be kept for several days in water, or wrapped in plastic wrapping paper and kept indefinitely at refrigerator temperature.

e. ORGANISMS STUDIED

We have used this method in mammalian erythrocytes, *Drosophila*, wheat seeds and leaves, and for certain other plants (see Table II).

f. COMMENT

It should be noted that the study of ICD can be conveniently combined with the study of other enzymes described in this chapter which use the same buffer system. These include AK, alkaline phosphatase, α-GPD, GA-3-PD, SDH, LAP, peroxidase, fumarase, aldolase, PK, and CK.

ICD also occurs in a DPN-dependent form in some lower species, such as yeast, and then generally requires AMP as a cofactor. Most animal tissues have the TPN-dependent form. Generally, study of the isozyme is best at a pH around neutrality, since some ICD isozymes are unstable at alkaline pH. The presence of three or four isozymes has been reported in rat and human tissues and in human serum (Tsao, 1960; Bell and Baron, 1962; Campbell and Maas, 1962). A fast isozyme predominates in the liver, whereas a slower isozyme predominates in the heart. The slower isozyme is much less stable, accounting for the failure of ICD to be detected in the serum after myocardial infarction. The isozymes of ICD have also been investigated with respect to intracellular localization (Lowenstein and Smith, 1962). Genetic variants of hepatic ICD have been found in mice (Henderson, 1965).

16. Succinate Dehydrogenase (SDH)

a. REFERENCES

The application of this buffer system to SDH was developed in our laboratory. For additional methodology see Hirsch *et al.* (1963) and Rossi *et al.* (1964).

b. BASIS FOR THE STAINING PROCEDURE

The site of SDH isozymes are stained according to the following reactions:

$$\text{Succinate} + \text{DPN} \xrightarrow{\text{SDH}} \text{fumarate} + \text{DPNH}$$

Presumably SDH activity generates DPNH. The site of DPNH production is then marked through the tetrazolium system (Section C, 2). ATP presumably acts as cofactor.

c. Reagents

(1) Gel buffer
 0.005 M histidine
The pH is adjusted to 7.0 with 2 N sodium hydroxide.
(2) Bridge buffer
 0.41 M sodium citrate
The pH is adjusted to 7.0 with 0.41 M citric acid.
(3) Staining mixture
 0.0154 M sodium succinate
 0.001 M DPN
 0.00091 M ATP
 0.000163 M phenazine methosulfate
 0.00043 M NB tetrazolium
 0.01 M disodium EDTA
 0.05 M K_2HPO_4, pH 7.0

The succinate and EDTA are added to the potassium phosphate buffer and then the other reagents added. The pH is readjusted to 7.0 as necessary with HCl.

d. Procedure

The gel is prepared and set up in the usual way. Electrophoresis is carried out for 4 hours at 8–10 V per linear centimeter of gel at approximately 15–20 mA per gel. After completion of electrophoresis the gel is sliced and stained in the usual manner. At the end of incubation the staining mixture may be removed and replaced with 50% ethanol. After 24 hours of exposure to ethanol the gels may be kept for several days in water, or wrapped in plastic wrapping paper.

e. Organisms Studied

We have used this method for the study of mammalian liver and heart (Table II).

f. Comment

It should be noted that the study of SDH can be conveniently combined with the study of the other enzymes described in this chapter which use the same buffer system. These include AK, alkaline phosphatase, α-GPD, GA-3-PD, ICD, LAP, peroxidase, fumarase, aldolase, PK, and CK.

17. Leucine Aminopeptidase (LAP)

a. REFERENCES

This is an unpublished method developed in our laboratory. The staining mixture is modified after Beckman and Johnson (1964).

b. BASIS FOR THE STAINING PROCEDURE

The site of LAP isozymes are stained by the cleavage of the peptide linkage of L-leucyl-β-naphthylamide HCl by the action of LAP, followed by coupling of the liberated naphthylamine to a dye such as Fast Black K.

c. REAGENTS

(1) Gel buffer
 0.005 M histidine
The pH is adjusted to 7.0 with 2 N sodium hydroxide.
(2) Bridge buffer
 0.41 M sodium citrate
The pH is adjusted to 7.0 with 0.41 M citric acid.
(3) Staining mixture
 0.00068 M L-leucyl-β-naphthylamide HCl (Sigma Chemical Co., St. Louis, Missouri)
 50 mg Fast Black K
 0.02 M tris adjusted to pH 5.2 with 1 N sodium hydroxide
 0.02 M malic acid adjusted to pH 5.2 with 1 N sodium hydroxide

The L-leucyl-β-naphthylamide must be the HCl salt and not the free base, because the latter is not soluble in tris-malic acid buffer. Prior to addition of the staining mixture the gels are preincubated at 5°C in a 0.5 M boric acid solution containing 0.005 M magnesium chloride for 1 hour. After rinsing with water, the gels are then ready for staining with the above staining mixture.

d. PROCEDURE

The gel is prepared and set up in the usual way. Electrophoresis is carried out for 4 hours at 8–10 V per linear centimeter of gel and approximately 15–20 mA per gel. After completion of electrophoresis the gel is sliced in the usual manner. Note, however, as indicated in Section E, 17, c, (3), that the gel is preincubated for 1 hour in a 0.5 M boric acid and 0.005 M magnesium chloride solution prior to staining. After rinsing the gel is stained in the usual manner.

e. ORGANISMS STUDIED

We have used this method for *Drosophila*, wheat leaves, and for a variety of other plants (see Table II). However, the method does not work satisfactorily with erythrocytes and an alternative method must be used (see below).

f. LAP METHOD FOR ERYTHROCYTES

This method was developed by Beckman and Johnson (1964). The staining mixture is the same as above [Section E, 17, c, (3)].
(1) Gel and bridge buffers
   0.1 M tris
   0.1 M boric acid
   0.005 M magnesium chloride
The pH is adjusted to 8.9 with 4 N sodium hydroxide.
(2) Procedure
The gel is prepared and set up in the usual way. Electrophoresis is carried out for 18–20 hours at approximately 10–12 V per linear centimeter of gel and approximately 5–10 mA per gel. After completion of electrophoresis the gel is sliced and stained as described in Section E, 17, c, (3). Note the preincubation step.

g. COMMENT

It should be noted that the study of LAP with the histidine gel buffer method (first method presented) can be conveniently combined with the study of other enzymes described in this chapter using the same buffer system. These include AK, alkaline phosphatase, α-GPD, GA-3-PD, ICD, SDH, peroxidase, fumarase, aldolase, PK, and CK.

LAP is an exopeptidase which is operationally defined as an enzyme that hydrolyzes the substrate L-leucyl-β-naphthylamide HCl. However, the enzyme is usually capable of a broader exopeptidase activity and it has been suggested (Scandalios, 1969) that the broader term "aminopeptidase" would be more appropriate. The term "arylamidase" has also been used.

LAP is widely distributed in plants (reviewed by Scandalios, 1969) and in animals (reviewed by Latner and Skillen, 1968; Wilkinson, 1965; Beckman *et al.*, 1966a).

LAP isozymes are present in human and animal sera (Lawrence *et al.*, 1960; Kowlessar *et al.*, 1960; Monis, 1964, 1965). Pineda *et al.* (1960) suggest that the liver is the major source of human serum LAP, while Monis (1964, 1965) suggests that the rat serum LAP arises from dermal fibroblasts and a urinary isozyme from renal tubular cells. Smith and

Rutenberg (1963, 1966) report tissue specific patterns in many tissues of the human and some variation in inhibitory properties in response to L-methionine. Their data suggest that the liver is the major source of the serum enzyme, a conclusion also reached by Beckman *et al.* (1966a).

Studies of LAP isozymes have been carried out in a number of disease states, mostly however, consisting of simple observation of electrophoretic migration (Kowlessar *et al.*, 1960; Beckman *et al.*, 1966a; Scandalios, 1967). Many differences have been observed particularly in liver diseases. This system may be of considerable clinical importance and further studies are indicated including additional characterization of the isozymes beyond electrophoretic migration.

Beckman *et al.* (1966b) have described four LAP isozymes in placental extracts and four serum LAP isozymes in pregnancy, although according to Beckman *et al.* (1966a,b) no correspondence was found between the two sets of isozymes. Pregnancy LAP isozymes have also been studied by Scandalios (1967) who finds that the first pregnancy serum LAP band corresponds in time to the period when placental development is becoming marked.

Scandalios (1969) has intensively studied and reviewed the LAP isozymes of maize.

18. Peroxidase

a. REFERENCES

The application of this buffer system to peroxidase was developed in our laboratory. For additional references and methods see Shannon (1968) and Scandalios (1969).

b. BASIS FOR THE STAINING PROCEDURE

The sites of peroxidase isozymes are stained by the following reaction:

$$2H_2O_2 \xrightarrow{\text{peroxidase}} 2 H_2O + O_2$$

The liberated oxygen then oxidizes a colorless compound, in this case benzidine, to a colored compound.

c. REAGENTS

(1) Gel buffer
    0.005 M histidine
The pH is adjusted to 7.0 with 2 N sodium hydroxide.
(2) Bridge buffer
    0.41 M sodium citrate

The pH is adjusted to 7.0 with 0.41 M citric acid.
  (3) Staining mixture
    0.0024 M benzidine (45 mg/100 ml of staining mixture)
    50% ethyl alcohol, final concentration; 50 ml of absolute alcohol/100 ml of staining mixture
    1% hydrogen peroxide, final concentration; 33 ml of 3% hydrogen peroxide/100 ml of staining mixture
    5% acetic acid, final concentration; 5 ml of glacial acetic acid/100 ml of staining mixture
    water, bring to 100 ml
The staining mixture should be made up immediately before use since hydrogen peroxide is not very stable.

d. Procedure

The gel is prepared and set up in the usual way. Electrophoresis is carried out for 4 hours at approximately 8–10 V per linear centimeter of gel and about 15–20 mA per gel. After completion of electrophoresis the gel is sliced and stained in the usual manner. The gel should be observed promptly after the staining mixture is added because the bands disappear very soon after addition of the staining mixture.

e. Organisms Studied

This method has been used in our laboratory for the study of germinating wheat seeds and for wheat leaves.

f. Comment

It should be noted that the study of peroxidase can be conveniently combined with the study of other enzymes described in this chapter in which the same buffer system is used. These include AK, alkaline phosphatase, α-GPD, GA-3-PD, ICD, SDH, LAP, fumarase, aldolase, PK, and CK.

The study of peroxidase isozymes has been most prominent in plants, where it has probably been studied more than any other isozyme system. Shannon (1968) in his review of plant isozymes lists 41 publications dealing with peroxidase isozymes. Plant peroxidase isozymes have also been reviewed by Scandalios (1969). Part of the great interest in plant peroxidases stems from the probable role of these enzymes in the oxidation of the plant hormone, indoleacetic acid. A great number of peroxidase isozymes occur in many plant tissues. In addition, there is considerable tissue and ontogenetic variation in bands and activity, and also considerable genetic variation in this system. This system deserves the fullest investigation; it is potentially of the greatest physiological importance in plants and, apparently, rich in variation and flexibility. It may be that much of importance in the areas of hormone action, including mechanisms of tissue differentiation and development, will be learned. It is quite likely that in systems such as this, the various roles of isozymes will eventually be revealed.

19. Fumarase

a. REFERENCES

This is an unpublished method developed in our laboratory.

b. BASIS FOR THE STAINING PROCEDURE

The sites of fumarase isozymes are stained according to the following reactions:

$$\text{Fumarate} \xrightarrow{\text{fumarase}} \text{malate}$$

$$\text{Malate} + \text{DPN} \xrightarrow{\text{MDH}} \text{oxalacetate} + \text{DPNH}$$

Fumarase activity is coupled to MDH activity which produces DPNH. The sites of DPNH generation are marked with the tetrazolium system (see Section C, 2).

c. REAGENTS

(1) Gel buffer
    0.005 M histidine
The pH is adjusted to 7.0 with 2 N sodium hydroxide.
(2) Bridge buffer
    0.41 M sodium citrate
The pH is adjusted to 7.0 with 0.41 M citric acid.
(3) Staining mixture
    0.1 M fumaric acid
    0.0009 M DPN
    0.2 mg malic dehydrogenase/100 ml staining solution
    0.00033 M phenazine methosulfate
    0.00042 M NB tetrazolium
    0.02 M sodium phosphate (dibasic), pH 7.0

d. PROCEDURE

The gel is prepared and set up in the usual manner. Electrophoresis is carried out for 4 hours at 8–10 V per linear centimeter of gel with about 15–20 mA per gel. After completion of electrophoresis the gel is sliced and stained in the usual manner.

e. ORGANISMS STUDIED

This method has been used for the study of mammalian liver and heart, *Drosophila*, and wheat seeds (see Table II).

f. COMMENT

The study of fumarase isozymes can be conveniently combined with other methods using the same buffer system. These include AK, alkaline phosphatase, α-GPD, GA-3-PD, ICD, SDH, LAP, peroxidase, aldolase, PK, and CK.

To date this system has not received much attention.

20. Aldolase

a. REFERENCES

The application of this buffer system to aldolase was developed in our laboratory. For additional references and methods see Anstall *et al.* (1966).

b. BASIS FOR THE STAINING PROCEDURE

The sites of aldolase isozymes are stained according to the following reactions:

$$\text{F-1,6-DP} \xrightarrow{\text{aldolase}} \text{GA-3-P} + \text{DHAP}$$

$$\text{GA-3-P} + \text{DPN} \xrightarrow{\text{GA-3-PD}} \text{1,3-DPG} + \text{DPNH}$$

Aldolase activity is coupled to GA-3-PD, which generates DPNH. The sites of DPNH generation are marked with the tetrazolium system (see Section C, 2).

c. REAGENTS

(1) Gel buffer
  0.005 M histidine
The pH is adjusted to 7.0 and 2.0 N sodium hydroxide.
(2) Bridge buffer
  0.41 M sodium citrate
The pH is adjusted to 7.0 with 0.41 M citric acid.
(3) Staining mixture
  0.007 M fructose 1,6-diphosphate

0.00075 M DPN
0.67 mg GA-3-PD/100 ml staining solution
0.00036 M NB tetrazolium
0.00033 M phenazine methosulfate
0.05 M tris, pH 7.1

d. PROCEDURE

The gel is prepared and set up in the usual manner. Electrophoresis is carried out for 4 hours at 8–10 V per linear centimeter of gel with about 15–20 mA per gel. After completion of electrophoresis the gel is sliced and stained in the usual manner.

e. ORGANISMS STUDIED

We have used this method for the study of mammalian erythrocytes (see Table II).

f. COMMENT

The study of aldolase isozymes can be conveniently combined with other methods using the same buffer system. These include AK, alkaline phosphatase, α-GPD, GA-3-PD, ICD, SDH, LAP, peroxidase, fumarase, PK, CK.

This isozyme system has not seen widespread study. Isozymes have been reported in human, rat, and frog tissues, with tissue differences (Anstall et al., 1966).

21. Pyruvate Kinase (PK)

a. REFERENCES

The application of this buffer system to PK was developed in our laboratory. For additional references and methods see Koler et al. (1964) and Tanaka et al. (1965).

b. BASIS FOR THE STAINING PROCEDURE

$$ADP + PEP \xrightarrow{PK} ATP + pyruvate$$

$$ATP + glucose \xrightarrow{hexokinase} ADP + G\text{-}6\text{-}P$$

$$G\text{-}6\text{-}P + TPN \xrightarrow{G\text{-}6\text{-}PD} 6\text{-}PG + TPNH$$

PK activity is coupled to hexokinase and G-6-PD, which produces TPNH. The sites of TPNH generation are marked with the tetrazolium system (see Section C, 2).

c. Reagents

(1) Gel buffer
  0.005 M histidine
The pH is adjusted to 7.0 with 2 N sodium hydroxide.
(2) Bridge buffer
  0.41 M sodium citrate
The pH is adjusted to 7.0 with 0.41 M citric acid.
(3) Staining mixture
  0.0015 M phospho-(enol)pyruvic acid
  0.0011 M ADP
  0.01 M glucose
  0.011 M MgCl$_2$
  0.00043 M TPN
  0.5 mg G-6-PD/100 ml staining solution
  0.1 mg hexokinase/100 ml staining solution
  0.00039 M phenazine methosulfate
  0.00029 M MTT tetrazolium
  0.75 gm Ionagar/100 ml staining solution
  0.1 M tris pH 8.0

Because of the double coupling to detect the bands, with soluble intermediates, an agar overlay method is used. The Ionagar is dissolved in ½ to ¾ of the tris buffer and heated to boiling. This mixture is then allowed to cool to 45°C. While it is cooling, the other reagents of the staining mixture are added to the remaining tris buffer and this mixture stirred thoroughly until all constituents are dissolved. The two mixtures are then added together and stirred. This solution is now ready to pour as an agar overlay.

d. Procedure

The gel is prepared and set up in the usual manner. Electrophoresis is carried out for 4 hours at 8–10 V per linear centimeter of gel with about 15–20 mA per gel. After completion of electrophoresis the gel is sliced in the usual manner and the staining mixture containing agar is prepared as described in the preceding section. The liquid staining mixture (temperature 40°–45°C) is poured over the sliced gel in a staining tray. As the agar solution cools to room temperature it will solidify. The gels may then be incubated in an incubator at 37°C for 2 hours or until staining is complete.

e. Organisms Studied

This method has been applied to the study of mammalian erythrocytes, liver, heart, wheat seeds, and *Drosophila* (see Table II). The specificity of the isozymes observed, except for *Drosophila*, is not clear, however, since AK isozymes also stain with this method (see the following section).

f. COMMENTS

The study of PK isozymes can be conveniently combined with other methods using the same buffer system. These include AK, alkaline phosphatase, α-GPD, GA-3-PD, ICD, SDH, LAP, peroxidase fumarase, aldolase, and CK.

It is necessary when using this method to stain one slice of the gel for AK by omitting the PEP substrate. This will allow the identification of AK isozymes which will also be present in the PK pattern. It is of considerable interest that the AK and PK patterns are identical in a number of tissues and species (see Table II). Since PK activity is known to be present in such tissues due to its role in glycolysis, such data would suggest dual specificity for at least one isozyme in each tissue. Such dual specificity may have evolutionary and indeed metabolic regulatory significance, since ADP-ATP interactions are central to intermediary metabolism. It may be possible in the future to devise methods to at least partially differentiate the two activities, by taking advantage of inhibitors, pH optima, etc. *Drosophila* has been the only organism to show a band which demonstrated PK, and not AK activity (see Table II).

Another method has been developed for PK involving coupling to LDH. Bands are demonstrated by an absence of fluorescence when the gel is exposed to ultraviolet light. Although this method is more specific, it has not been very satisfactory in our hands because the bands are not distinct, at least by visual observation.

22. Creatine Kinase (CK)

a. REFERENCES

The application of this buffer system to CK was developed in our laboratory. For additional methods and references see Rosalki (1965) and Eppenberger *et al.* (1967).

b. BASIS FOR THE STAINING PROCEDURE

The sites of CK isozymes are stained according to the following reactions:

$$\text{ADP} + \text{creatine-PO}_4 \xrightarrow{\text{CK}} \text{ATP} + \text{creatine}$$

E. Isozyme Methods Employed in Our Laboratory / 109

$$\text{ATP} + \text{glucose} \xrightarrow{\text{hexokinase}} \text{ADP} + \text{G-6-P}$$

$$\text{G-6-P} + \text{TPN} \xrightarrow{\text{G-6-PD}} \text{6-PG} + \text{TPNH}$$

CK activity is coupled to hexokinase and G-6-PD, which produces TPNH. The sites of TPNH generation are marked with the tetrazolium system (see Section C, 2).

c. REAGENTS

(1) Gel buffer
0.005 M histidine
The pH is adjusted to 7.0 with 2 N sodium hydroxide.
(2) Bridge buffer
0.41 M sodium citrate
The pH is adjusted to 7.0 with 0.41 M citric acid.
(3) Staining mixture
0.0039 M phosphocreatine
0.0011 M ADP
0.01 M glucose
0.011 M $MgCl_2$
0.00043 M TPN
0.1 mg hexokinase/100 ml staining solution
0.5 mg G-6-PD
0.000392 M phenazine methosulfate
0.00029 M MTT tetrazolium
0.75 gm Ionagar/100 ml staining solution
0.1 M tris, pH 8.0

Because of the double coupling to detect the bands, with soluble intermediates, an agar overlay method is used. The Ionagar is dissolved in ½ to ¾ of the tris buffer and heated to boiling. This mixture is then allowed to cool to 45°C. While it is cooling, the other reagents of the staining mixture are added to the remaining tris buffer and this mixture is stirred thoroughly until all constituents are dissolved. The two mixtures are then added together and stirred. This solution is now ready to pour as an agar overlay.

d. PROCEDURE

The gel is prepared and set up in the usual manner. Electrophoresis is carried out for 4 hours at 8–10 V per linear centimeter of gel with about 15–20 mA per gel. After completion of electrophoresis the gel is sliced in the usual manner and the staining mixture containing agar is prepared as described in the preceding section. The liquid staining mixture (temperature 40°–45°C) is poured over the sliced gel in a staining tray. As the agar solution cools to room temperature it will solidify. The gels may then be incubated in an incubator at 37°C for 2 hours or until staining is complete.

e. ORGANISMS STUDIED

This method has been applied to the study of mammalian erythrocytes, liver, heart, skeletal muscle, wheat seeds, *Drosophila,* and commercial enzyme (see Table II). The specificity of the isozymes observed, except for the skeletal muscle and commercial enzyme, is not clear however, since AK isozymes also stain with this method (see the following section).

f. COMMENT

The study of CK isozymes can be conveniently combined with other methods using the same buffer system. These include AK, alkaline phosphatase, α-GPD, GA-3-PD, ICD, SDH, LAP, peroxidase, fumarase, aldolase, and PK.

It is necessary when using this method to stain one slice of the gel for AK by omitting creatine phosphate substrate. This will allow the identification of AK isozymes which will also be present in the CK pattern. The ability of this method to identify CK activity has been shown by electrophoresis of skeletal muscle and commercial enzyme. In the various species and tissues studied however, CK and AK patterns were usually identical. This may mean that one or more isozymes have dual specificity, although this inference cannot be drawn unless it is certain that CK activity is present in the tissue studied. It may be possible in the future to devise methods to at least partially differentiate the two activities, by taking advantage of inhibitors, pH optima, etc.

Considerable work has been done in birds, which show some differences in patterns (Eppenberger *et al.,* 1967). Rosalki (1965) has studied CK isozymes in human skeletal and cardiac muscle. Different patterns were observed between heart muscle and red skeletal muscle, on the one hand, and white skeletal muscle, on the other.

Another method has been developed for CK involving coupling to PK and LDH. Bands are demonstrated as an absence of fluorescence when the gel is exposed to ultraviolet light. Although this method has greater specificity, we have not obtained very satisfactory results with it because the bands are not distinct, at least by visual observation.

23. Catalase

a. REFERENCES

The application of this buffer system to catalase was developed in our laboratory. The staining method was developed by Thorup *et al.* (1961).

b. BASIS FOR THE STAINING PROCEDURE

The sites of catalase isozymes are stained according to the following reactions:

$$2H_2O_2 \xrightarrow{\text{catalase}} 2H_2O + O_2$$

$$H_2O_2 + I^- \longrightarrow I_2$$

This is a two-step procedure. The gel is first immersed for a few seconds in a solution of sodium thiosulfate and hydrogen peroxide. The action of catalase at the site of its isozymes destroys hydrogen peroxide, but hydrogen peroxide is present elsewhere on the gel. Upon exposure to potassium iodide, wherever hydrogen peroxide is present, the iodide is oxidized to iodine, which turns the gel blue-black. The gel remains colorless in the areas of catalase activity. The thiosulfate in the staining solution is inactivated by hydrogen peroxide except in the areas of catalase activity. Its function there is to reduce any iodine which escapes into the solution and settles on the catalase area (Thorup *et al.*, 1961).

c. REAGENTS

(1) Gel buffer
    0.005 M histidine
The pH is adjusted to 8.0 with 2 N sodium hydroxide.
(2) Bridge buffer
    0.41 M sodium citrate
The pH is adjusted to 8.0 with 0.41 M citric acid.
(3) Staining mixture
Note that there are two parts to the staining solution which are not mixed.
    Solution 1
        30 ml of 0.06 M sodium thiosulfate
        70 ml of 3% H_2O_2
Do not mix until ready to use (see the following section).
    Solution 2
        100 ml of 0.09 M potassium iodide

See procedure in the following section for the use of these staining solutions.

d. Procedure

The gel is prepared and set up in the usual way. Electrophoresis is carried out for 4 hours at approximately 8–10 V per linear centimeter of gel and about 15–20 mA per gel. After completion of electrophoresis, the gel is sliced in the usual manner. The gel is allowed to remain in the staining dish for a few moments to warm up to room temperature. Then staining solution 1 is prepared by pouring the sodium thiosulfate and hydrogen peroxide together in the staining tray. The gel is immediately immersed in this staining solution for 30–60 seconds. The gel is then removed from solution 1 and placed into a second staining tray which contains staining solution 2 (potassium iodide). The staining tray is agitated by gentle rocking and the entire starch gel begins to turn a bluish-black color, except at the site of the catalase bands, which remain clear. Photographs or drawings must be made immediately since the bands are transitory.

e. Organisms Studied

We have used this method for mammalian erythrocytes and wheat leaves.

f. Comment

It should be noted that the study of catalase can be conveniently combined with the study of other enzymes described in this chapter which use the same buffer system. These include GR, MDH, ADH, diaphorase, TPI, acetylcholinesterase, GDH, and XDH.

Catalase is widely distributed in the plant and animal kingdom. Its specific role is unknown, although it is generally regarded as an enzyme which protects against hydrogen peroxide toxicity. It is the most active enzyme known; one catalase molecule is capable of destroying 44,000 molecules of hydrogen peroxide per second (Thorup et al., 1961).

The status of catalase isozymes in maize has been well reviewed by Scandalios (1969). Considerable genetic, tissue, and ontogenetic variation is present.

A few studies have been done in animals. Most studies showing isozymes have used chromatography. More than one band of activity has been reported in human and rat liver and erythrocytes (Nishimura et al., 1964).

An interesting clinical condition called acatalasia occurs in man. The condition is autosomal recessive, and is characterized by very low catalase levels in all tissues studied. The disease occurs in both the Japanese and the Swiss populations, and appears to be due to different genes in the two cases. Erythrocyte catalase has been recently reviewed by Aebi and Suter (1969).

24. Glutathione Reductase (GR)

a. REFERENCE

This is a method developed by Brewer and Sing (1969).

b. BASIS FOR THE STAINING PROCEDURE

The sites of GR isozymes are stained according to the following reaction:

$$GSSG + TPNH \xrightarrow{GR} GSH + TPN$$

GR activity generates GSH. The site of GSH production is marked through the production of a colored compound by its action on dithiobis.

c. REAGENTS

(1) Gel buffer
  0.005 M histidine
The pH is adjusted to 8.0 with 2 N sodium hydroxide.
(2) Bridge buffer
  0.41 M sodium citrate
The pH is adjusted to 8.0 with 0.41 M citric acid.
(3) Staining mixture
  0.0055 M GSSG
  0.00032 M TPNH
  0.00088 M 2-nitrobenzoic acid (synonym, dithiobis)
  0.033 M EDTA
  1 gm of Ionagar/100 ml staining solution
  0.133 M tris, pH 8.0

The EDTA and 2-nitrobenzoic acid are added to half of the tris buffer and heated only as much as necessary to bring the nitrobenzoic acid into solution. When the nitrobenzoic acid has been completely dissolved the solution is allowed to cool. After the mixture reaches a temperature of 45°C the TPNH and GSSG may be added. The agar is added to the remaining half of the tris buffer and heated to boiling. After the agar solution has cooled to 45°C the two solutions are mixed and stirred well. This solution is then used as an agar overlay as described in the next section.

d. PROCEDURE

The gel is prepared and set up in the usual way. Electrophoresis is carried out for 4 hours at approximately 8–10 V per linear centimeter of gel with about 15–20 mA per gel. After completion of electrophoresis the gel is sliced in the usual manner. The agar-containing staining mixture at approximately 40°–45°C is poured over the cut surface of the gel. As it cools to room temperature the staining mixture solidifies. Gels may then be incubated at 37°C for 1–2 hours, or as long as necessary to bring out the yellowish bands marking the sites of glutathione reductase.

e. ORGANISMS STUDIED

We have used this method for the study of mammalian erythrocytes (see Table II).

f. COMMENT

It should be noted that the study of GR can be conveniently combined with the study of other enzymes using the same buffer system. These include catalase, MDH, ADH, diaphorase, TPI, acetylcholinesterase, GDH, and XDH.

With the dithiobis method, three yellowish bands appear with human erythrocytes. Only the most cathodal of these is specific for glutathione reductase since the other two appear when oxidized glutathione is omitted from the staining mixture. Similar results have been obtained with other methods. Thus, particularly in the case of glutathione reductase it is important to do control studies in which one half of the gel is stained with the staining mixture from which oxidized glutathione is absent.

An alternative staining method (Kaplan, 1968) may be used which produces bluish bands. This staining mixture is as follows:

$0.0034\ M$ GSSG
$0.00036\ M$ TPNH
$0.0011\ M$ MTT tetrazolium
$0.000052\ M$ DCIP
$0.25\ M$ tris, pH 8.4

This mixture can be used in solution without an agar overlay.

Long (1966, 1967) has reported a GR method which produces negative staining. Long (1967) has also found electrophoretic variation in the American Negro. Inheritance was autosomal. He also reports an interesting association between the electrophoretically fast variant and primary gout. Erythrocyte GR has been recently reviewed by Brewer (1969).

25. Malate Dehydrogenase (MDH)

a. REFERENCES

The application of this buffer system to MDH was developed in our laboratory (Brewer and Sing, 1969). For references and methods see Markert and Moller (1959), Tsao (1960), and Latner and Skillen (1962).

b. BASIS FOR THE STAINING PROCEDURE

The sites of MDH isozymes are stained according to the following reactions:

$$\text{Malate} + \text{DPN} \xrightarrow{\text{MDH}} \text{oxalacetate} + \text{DPNH}$$

MDH activity generates DPNH. The site of DPNH production is marked through the tetrazolium system.

c. REAGENTS

(1) Gel buffer
  0.005 M histidine
The pH is adjusted to 8.0 with 2 N sodium hydroxide.
(2) Bridge buffer
  0.41 M sodium citrate
The pH is adjusted to 8.0 with 0.41 M citric acid.
  Best results are obtained if 0.00015 M DPN is added to the cathodal bridge tray (not the electrode compartment).
(3) Staining mixture
  0.2 M DL-malic acid
  0.001 M DPN
  0.000163 M phenazine methosulfate
  0.00043 M NB tetrazolium
  0.05 M tris, pH 7.0
It is convenient to mix the tris and malic acid and adjust the pH to 7.0 with 4 N sodium hydroxide, and then add the other reagents. The mixture is strained through glass wool before applying to the gel.

d. PROCEDURE

0.00015 M DPN is added to the molten gel before degassing. Aside from this, the gel is prepared and set up in the usual way. Electrophoresis is carried out for 4 hours at approximately 8–10 V per linear centimeter of gel, and about 15–20 mA per gel. After completion of electrophoresis, the gel is sliced and stained in the usual manner. At the end of incubation the staining mixture may be removed and replaced with 50% ethanol. After 24 hours of exposure to ethanol, the gels may be kept for several days in water, or wrapped in plastic wrapping paper and kept indefinitely at refrigerator temperatures.

e. ORGANISMS STUDIED

We have used this method for mammalian erythrocytes, *Drosophila*, wheat seeds and leaves, and for several other plants (see Table II).

f. COMMENT

It should be noted that the study of MDH can be conveniently combined with the study of other enzymes described in this chapter which use the same buffer system. These include catalase, GR, ADH, diaphorase, TPI, acetylcholinesterase, GDH, and XDH.

Human serum MDH was separated by Vesell and Bearn (1958) into three fractions by starch block electrophoresis. Heterogeneity of this enzyme in various animal tissue was subsequently reported by a number of investigators including Markert and Moller (1959), Tsao (1960), Latner and Skillen (1962), Yakulis *et al.* (1962), Goldberg (1963), and others.

The enzyme has some fascinating aspects. It is one of a group of enzymes known to exist in distinct forms in the soluble cytoplasm and in the mitochondria. The cytoplasmic and mitochondrial types of isozymes differ in electrophoretic mobility (Wieland *et al.*, 1959), and in certain kinetic characteristics (Delbruck *et al.*, 1959; Thorne, 1960; Kaplan and Ciotti, 1961). It has been observed that the mitochondrial enzyme preferentially oxidizes malate, while the soluble enzyme preferentially catalyzes the reverse reaction. Kaplan (1961, 1963) concludes that this relationship leads to ATP production from the DPNH produced in the mitochondria. The oxalacetate diffuses into the cytoplasm where it is reduced to malate by the cytoplasmic enzyme. Diffusion of the malate into the mitochondria allows repetition of the cycle. This scheme is similar to that proposed by Bucher and Klingenburg (1958) and Sactor (1959) for the soluble and mitochondrial forms of α-GPD.

Davidson and Cortner (1967a,b) have screened for genetic variation in the two types of enzymes in human populations. They observed only one electrophoretic variant in the cytoplasmic enzyme in 3000 individuals (Davidson and Cortner, 1967a). Variation in the mitochondrial enzyme was more frequent, with approximately 1% of the population showing an electrophoretic variant form in their leukocytes. According to Davidson and Cortner (1967b) this is the first demonstration of a genetic variant of a mitochondrial protein in man. The variants showed typical Mendelian segregation (probably autosomal), rather than maternal inheritance, which is different than findings from certain lower organisms. In *Neurospera* for example, mutations of mitochondrial DNA have been observed which are transmitted only by the maternal parent (Luck and Reich, 1964; Reich and Luck, 1966), supporting the concept that mitochondria have a separate genetic system. Davidson and Cortner (1967b) suggest that some mitochondrial proteins are under nuclear (chromosomal) genetic control, while others are under mitochondrial DNA control.

Recent studies by Kitto et al. (1967) in *Neurospera* indicate that mitochondrial and cytoplasmic MDH and aspartate aminotransferase represent four separate physical entities, in contrast to a suggestion by Munkres (1965), who suggested that MDH and aspartate aminotransferase activities resided in the same protein.

Shows and Ruddle (1968) have reported electrophoretic studies of TPN-MDH, which catalyzes the oxidative decarboxylation of malate in the presence of TPN and Mn^{++} to pyruvate and CO_2. A cytoplasmic and a mitochondrial form occurs in the mouse. The cytoplasmic form shows genetic variation in *Mus musculus*. The relationship of these enzymes to DPN-MDH is not clear.

26. Alcohol Dehydrogenase (ADH)

a. Reference

The application of this buffer system to ADH was developed in our laboratory. For additional references and methods see Ursprung and Leone (1965) and Scandalios (1969).

b. Basis for the Staining Procedure

The sites of ADH isozymes are stained according to the following reactions:

$$RCH_2OH + DPN \xrightarrow{ADH} RCHO + DPNH$$

(In the case of ethyl alcohol, R is a methyl group.)

ADH activity generates DPNH. The site of DPNH production is marked through the tetrazolium system (Section C, 2).

c. Reagents

(1) Gel buffer
    0.005 M histidine

The pH is adjusted to 8.0 with 2 N sodium hydroxide.
(2) Bridge buffer
   0.41 M sodium citrate
The pH is adjusted to 8.0 with 0.41 M citric acid.
(3) Staining mixture
   7.5 ml of 95% ethyl alcohol/100 ml staining solution
   0.001 M DPN
   0.000163 M phenazine methosulfate
   0.00043 M NB tetrazolium
   0.05 M sodium phosphate (dibasic), pH 7.0

The staining mixture should be made up shortly before use since many of the reagents are not very stable.

d. PROCEDURE

The addition of 10 mg of DPN to the molten gel before degassing may improve results. Other than this, the gel is prepared and set up in the usual way. Electrophoresis is carried out for 4 hours at approximately 8–10 V per linear centimeter of gel and about 15–20 mA per gel. After completion of electrophoresis the gel is sliced and stained in the usual manner.

e. ORGANISMS STUDIED

We have used this method for the study of mammalian liver, *Drosophila*, wheat, and several other plants (see Table II).

f. COMMENT

It should be noted that the study of ADH can be conveniently combined with the study of other enzymes described in this chapter which use the same buffer system. These include catalase, GR, MDH, diaphorase, TPI, acetylcholinesterase, GDH, and XDH.

ADH is widely distributed in the plant and animal kingdom. In general, it has broad specificity, reacting with a large number of primary and secondary, straight and branched chain, aliphatic and aromatic alcohols. For this reason its precise physiological functions are not known.

The first detailed genetic studies of electrophoretic variation in ADH were done in *Drosophila* (Ursprung and Leone, 1965; Grell et al., 1965). Considerable variation is present in plants (reviewed by Scandalios, 1969). Two closely linked loci, each with two alleles, controls ADH production in maize (Scandalios, 1969).

In general, mammalian ADH migrates cathodally. In certain mammalian tissues, particularly the liver, ADH has the interesting property of showing bands in the absence of substrate, leading to the term "nothing dehydrogenase" (Koen and Shaw, 1965; Shaw and Koen, 1965). As this system illustrates, the possibility of obtaining stained bands in the absence of substrate should always be borne in mind in the study of new systems, and one slice of the gel stained in the absence of substrate, in order to avoid confusion and errors.

Ressler and Stitzer (1967) suggest that the nothing dehydrogenase activity may be due to bound substrate. As they point out, other dehydrogenase activities (such as lactate) have also been detected in the absence of added substrate, with binding of substrate the most likely explanation. Ressler and Stitzer observe that nothing dehydrogenase is converted to an anodally migrating GDH during storage of liver tissue. They also presented evidence to suggest that the cathodal nothing dehydrogenase band can also liberate mitochondrial MDH.

27. Diaphorase (DPNH-Methemoglobin Reductase)

a. Reference

This method was developed by Brewer et al. (1967c).

b. Basis for the Staining Procedure

The sites of diaphorase isozymes are stained according to the following reactions:

$$\text{Oxidized DCIP} + \text{DPNH} \xrightarrow{\text{diaphorase}} \text{reduced DCIP} + \text{DPN}$$

The oxidized DCIP causes a bluish background to the gel. At the sites of diaphorase isozymes the DCIP is reduced. The diaphorase bands appear as whitish areas in a bluish background of the gel.

c. Reagents

(1) Gel buffer
    0.005 M histidine
The pH is adjusted to 8.0 with 2 N sodium hydroxide.
(2) Bridge buffer
    0.41 M sodium citrate
The pH is adjusted to 8.0 with 0.41 M citric acid.
(3) Staining mixture
    0.00014 M 2,6-dichlorophenolindophenol (DCIP)
    0.000176 M DPNH

0.75 gm Ionagar/100 ml staining solution
0.05 M tris, pH 8.0

The Ionagar is dissolved in ½–¾ of the tris buffer and heated to boiling. This mixture is then allowed to cool to 45°C. While it is cooling, the other reagents of the staining mixture are added to the remaining tris buffer and this mixture stirred thoroughly until all constituents are dissolved. The two mixtures are then added together and stirred. This solution is now ready to pour as an agar overlay.

d. Procedure

The gel is prepared and set up in the usual way. Electrophoresis is carried out for 4 hours at approximately 8–10 V per linear centimeter of gel and about 15–20 mA per gel. After completion of electrophoresis the gel is sliced in the usual manner and the staining mixture containing agar is prepared as described in the preceding section. The liquid staining mixture (temperature 40°–45°C) is poured over the sliced gel in a staining tray. As the agar solution cools to room temperature it will solidify. The gels may then be incubated in an incubator at 37°C for 2 hours, or until the whitish bands appear. Pictures or drawings should be made soon thereafter because the bands begin to diffuse after a period of time.

Diaphorase bands of human erythrocytes begin to appear in about 15 minutes after incubation at 37°C. It is most convenient to interpret or read them at about 15–45 minutes. In this system human hemoglobin migrates slightly cathodally and the diaphorase bands migrate slightly anodally.

e. Organisms Studied

This method has been used for mammalian erythrocytes (Table II).

f. Comment

It should be noted that the study of diaphorase can be conveniently combined with the study of other enzymes described in this chapter using the same buffer system. These include catalase, GR, MDH, ADH, TPI, acetylcholinesterase, GDH, and XDH.

This system has been established primarily to study DPNH-methemoglobin reductase isozymes occurring in mammalian erythrocytes. These isozymes have diaphorase activity, which is used to demonstrate the isozymes histochemically. It is quite likely that the system will reveal other enzymes with diaphorase activity in other tissues and other organisms.

E. Isozyme Methods Employed in Our Laboratory / 121

Sheep erythrocytes show an autosomally determined genetic polymorphism in electrophoretic migration in this system (Brewer et al., 1967c). Human erythrocytes show three isozymes but so far no clear cut variation (Brewer et al., 1967c).

An autosomal recessive gene most frequent in certain Alaskan populations causes a deficiency of this enzyme, producing the clinical condition, congenital methemoglobinemia (Scott and Griffith, 1959).

28. Triosephosphate Isomerase (TPI)

a. REFERENCE

This is an unpublished method developed in our laboratory.

b. BASIS FOR THE STAINING PROCEDURE

$$DHAP \xrightarrow{TPI} GA\text{-}3\text{-}P$$

$$GA\text{-}3\text{-}P + DPN + \text{phosphate or arsenate} \xrightarrow{GA\text{-}3\text{-}PD} 1,3\text{-}DPG + DPNH$$

TPI activity is coupled to GA-3-PD activity which generates DPNH. The sites of DPNH production are marked through the tetrazolium system (Section C, 2).

c. REAGENTS

(1) Gel buffer
 0.005 M histidine
The pH is adjusted to 8.0 with 2 N sodium hydroxide.
 (2) Bridge buffer
 0.41 M sodium citrate
The pH is adjusted to 8.0 with 0.41 M citric acid.
 (3) Staining mixture
 0.001 M DHAP
 0.0015 M DPN
 0.008 M sodium arsenate
 0.000163 M phenazine methosulfate
 0.00036 M NB tetrazolium
 0.2 mg GA-3-PD/100 ml staining solution
 0.02 M tris, pH 8.0

d. PROCEDURE

The gel is prepared and set up in the usual way. Electrophoresis is carried out for 4 hours at 8–10 V per linear centimeter of gel and about

15–20 mA per gel. After completion of electrophoresis the gel is sliced and stained in the usual manner. After 24 hours exposure to 50% ethanol, the gels may be kept for several days in water or wrapped in plastic wrapping paper and kept indefinitely at refrigerator temperatures.

e. ORGANISMS STUDIED

We have used this method for the study of mammalian liver and heart, *Drosophila*, and wheat seeds (Table II).

f. COMMENT

This method may be conveniently combined with a number of other methods included in this chapter which use the same buffer system, including catalase, GR, MDH, ADH, diaphorase, acetylcholinesterase, GDH, and XDH.

29. Acetylcholinesterase and Pseudocholinesterase

a. REFERENCES

The staining method is a modification of methods developed by Gomori (1952), Bernsohn *et al.* (1961) and Shaw (C. R. Shaw, personal communication).

b. BASIS FOR THE STAINING PROCEDURE (ACETYLCHOLINESTERASE)

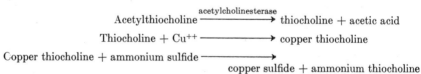

The acetylcholinesterase activity splits acetic acid from acetylthiocholine. The liberated thiocholine in the presence of copper ions forms a sparingly soluble white precipitate at the site of the activity. Upon the addition of ammonium sulfide, a brownish-black precipitate develops. The tetraisopropylphosphoramide is used as an inhibitor of other esterases.

c. REAGENTS (ACETYLCHOLINESTERASE)

(1) Gel buffer
  0.005 M histidine
The pH is adjusted to 8.0 with 2.0 N sodium hydroxide.
(2) Bridge buffer
  0.41 M sodium citrate

E. Isozyme Methods Employed in Our Laboratory / 123

The pH is adjusted to 8.0 with 0.41 M citric acid.
(3) Staining mixture
 (a) Incubation mixture
  2.4 M Na_2SO_4 (see note below)
  0.0069 M acetylthiocholine iodide
  0.006 M $CuSO_4$
  0.022 M glycine
  0.075 M maleic acid
  0.15 M NaOH
  0.025 M $MgCl_2$
  0.00001 M tetraisopropylphosphoramide (see note below)
Dissolve the Na_2SO_4 first, because the solution must be heated to effect solution. Then add the other reagents when the solution cools. The tetraisopropylphosphoramide is not soluble in water. It should be dissolved in 10 ml of 95% ethanol, which can then be added to the rest of the staining mixture. The pH is adjusted to 6.0. This mixture will be used for a 2 hour preliminary incubation of the gel (see procedure).
 (b) Developing mixture
  0.01 M $(NH_4)_2S$

d. Procedure

The gel is prepared and set up in the usual way. Electrophoresis is carried out for 4 hours at approximately 8–10 V per linear centimeter of gel, and about 15–20 mA per gel.

After completion of electrophoresis, the gel is sliced and the incubation mixture is poured over the gel which is allowed to incubate at 37°C for a period of 2 hours. After 2 hours, the gel is rinsed in 40% Na_2SO_4, then the developing mixture of $(NH_4)_2S$ is poured over the gel. Brown bands which appear quite rapidly denote enzymic activity.

e. Organisms Studied

This method has been used with mammalian erythrocytes, muscle, and commercial enzyme (see Table II).

f. Pseudocholinesterase

The method is the same as for acetylcholinesterase except that 1×10^{-6} M 1,5-bis-(4-trimethylammoniumphenyl)pentan-3-one diiodide is substituted for tetraisopropylphosphoramide in the staining mixture. This agent will inhibit "true" acetylcholinesterase, but will not inhibit pseudocholinesterase.

This method has been used with mammalian sera (Table II).

g. Comment

It should be noted that the study of cholinesterase isozymes can be conveniently combined with the study of other enzyme systems described in this chapter using the same buffer system. These include catalase, GR, MDH, ADH, diaphorase, TPI, GDH, and XDH.

Cholinesterases are thought to be confined to the animal kingdom. Their general property is the hydrolysis of cholinesters into free choline and the corresponding acid. The cholinesterases have been divided into two general types, so called "true" cholinesterase, and pseudocholinesterase. True cholinesterase has great activity toward acetylcholine and less activity toward other cholinesters. One physiological function of this enzyme is extremely important—the hydrolysis of acetylcholine at the neuromuscular function. A true cholinesterase is also firmly bound to the membrane of mammalian red cells. The function of the red cell enzyme is unknown. The general topic of red cell esterases has been reviewed by Tashian (1969).

Pseudocholinesterases have greatest activity on longer chain cholinesters such as butrylcholine and split acetylcholine relatively slowly. Pseudocholinesterases are found in many tissues including serum, brain, liver, heart, pancreas, and skin. Serum pseudocholinesterase is of clinical importance because it assists in the inactivation of certain drugs used in anesthesia. Most important in this regard is the muscle relaxant, suxamethonium, which is normally quite short-acting. However, an inherited deficiency of this condition, in homozygous form, causes prolonged apnea (failure to breathe) when this drug is given. The gene has an appreciable frequency in many populations such that homozygotes may occur as often as once in every 3000 individuals. Serum pseudocholinesterase has been nicely reviewed by Lehmann and Liddell (1964).

The occurrence of isozymes of cholinesterases in a variety of tissues and organisms has been recently reviewed by Latner and Skillen (1968).

30. Glutamate Dehydrogenase

a. References

The application of this buffer system to glutamate dehydrogenase was developed in our laboratory. For additional reference see Van derHelm (1962).

b. Basis for the Staining Procedure

The sites of glutamate dehydrogenase isozymes are stained according to the following reactions:

E. Isozyme Methods Employed in Our Laboratory 125

$$\text{Glutamate} + \text{DPN} + \text{H}_2\text{O} \xrightarrow{\text{GDH}} \alpha\text{-oxoglutarate} + \text{DPNH} + \text{NH}_4^+$$

Glutamate dehydrogenase activity generates DPNH. The sites of DPNH production are marked through the tetrazolium system (Section C, 2).

c. REAGENTS

(1) Gel buffer
    0.005 M histidine
The pH is adjusted to 8.0 with 2 N sodium hydroxide.
(2) Bridge buffer
    0.41 M sodium citrate
The pH is adjusted to 8.0 with 0.41 M citric acid.
(3) Staining mixture
    0.25 M L-glutamic acid
    0.0015 M DPN
    0.000163 M phenazine methosulfate
    0.00043 NB tetrazolium
    0.125 M sodium phosphate buffer (dibasic) pH 9.0
The staining mixture should be made up shortly before used since many of the reagents are not very stable.

d. PROCEDURE

The gel is prepared and set up in the usual way. Electrophoresis is carried out for 4 hours at about 8–10 V per linear centimeter of gel and about 15–20 mA per gel. After completion of electrophoresis, the gel is sliced and stained in the usual manner.

e. ORGANISMS STUDIED

This method has been used in our laboratory for the study of mammalian erythrocytes, liver, heart, muscle, *Drosophila*, and wheat seeds (Table II).

f. COMMENT

It should be noted that the study of GDH can be conveniently combined with other enzymes described in this chapter using the same buffer system. These include catalase, GR, MDH, ADH, diaphorase, TPI, acetylcholinesterase, and XDH.

The GDH of human tissues have been resolved by agar gel electrophoresis into five or six isozymes (Van derHelm, 1962)

31. Xanthine Dehydrogenase (XDH)

a. REFERENCE

This is an unpublished method developed in our laboratory.

b. BASIS FOR THE STAINING PROCEDURE

The sites of XDH isozymes are stained according to the following reactions:

$$\text{Hypoxanthine} + \text{DPN} \xrightarrow{\text{XDH}} \text{xanthine} + \text{DPNH}$$

$$\text{Xanthine} + \text{DPN} \xrightarrow{\text{XDH}} \text{uric acid} + \text{DPNH}$$

The activity of XDH generates DPNH. The sites of DPNH production are marked through the tetrazolium system (Section C, 2).

c. REAGENTS

(1) Gel buffer
   0.005 M histidine
The pH is adjusted to 8.0 with 2 N sodium hydroxide.
(2) Bridge buffer
   0.41 M sodium citrate
The pH is adjusted to 8.0 with 0.41 M citric acid.
(3) Staining mixture
   0.05 M hypoxanthine (see note below)
   0.001 M DPN
   0.000163 M phenazine methosulfate
   0.00043 M NB tetrazolium
   0.005 M tris, pH 7.0

Hypoxanthine is quite insoluble and should be heated in the tris buffer. The other reagents can be added when the solution cools.

d. PROCEDURE

The gel is prepared and set up in the usual way. Electrophoresis is carried out for 4 hours at approximately 8–10 V per linear centimeter of gel and about 15–20 mA per gel. After completion of electrophoresis the gel is sliced and stained in the usual manner.

e. ORGANISMS STUDIED

We have used this method for the study of *Drosophila* (Table II).

f. Comment

It should be noted that the study of XDH isozymes can be conveniently combined with the study of other enzymes described in this chapter using the same buffer system. These include catalase, GR, MDH, ADH, diaphorase, TPI, acetylcholinesterase, and GDH.

32. Acid Phosphatase

a. Reference

The application of this buffer system to acid phosphatase was developed in our laboratory. The staining system is that of Hopkinson et al. (1964).

b. Basis for the Staining Procedure

The sites of acid phosphatase isozymes are stained by virtue of the splitting of phosphate off phenolphthalein diphosphate at an acid pH. The liberated phenolphthalein turns a red color under the alkaline conditions of the addition of ammonium hydroxide to the staining mixture.

c. Reagents

(1) Gel buffer
   0.005 M histidine
The pH is adjusted to 6.0 with 2 N sodium hydroxide.
(2) Bridge buffer
   0.41 M sodium citrate
The pH is adjusted to 6.0 with 0.41 M citric acid.
(3) Staining mixture
   0.0075 M phenolphthalein diphosphate
   0.05 M citric acid, pH 6.0
The citric acid buffer is adjusted to pH 6.0 with 2 N sodium hydroxide before adding the phenolphthalein diphosphate and then the pH readjusted if necessary.

d. Procedure

The gel is prepared and set up in the usual way. Electrophoresis is carried out for 4 hours at approximately 8–10 V per linear centimeter of gel and about 15–20 mA per gel. After completion of electrophoresis the gel is sliced in the usual manner. The sliced gel is then incubated in the staining mixture for 3 hours at 37°C. Note that color will not develop at this point. The staining mixture is then removed and 2 ml of concentrated ammonium hydroxide added, in a hood. The staining tray is covered to make an ammonical atmosphere. Red zones which appear mark the site of acid phosphatase. By adding 50% ethanol immediately after appearance of the bands, diffusion can be reduced. The bands are very transitory so pictures or diagrams should be made within 15–20 minutes after addition of the ammonium hydroxide.

e. Organisms Studied

We have used this method for wheat and certain other plants (see Table II). This method does not work well with mammalian erythrocytes, and an alternative method should be used (see the following section).

f. Acid Phosphatase Method for Mammalian Erythrocytes

(1) Reference
This is the method of Hopkinson et al. (1964).
(2) Gel buffer
0.0046 M tris
0.0025 M succinic acid

The pH is adjusted to 6.0 by adding approximately 44 ml of 0.002 M succinic acid per liter of tris and succinic acid.

(3) Bridge buffer
0.41 M sodium citrate

The pH is adjusted to 6.0 by adding 0.41 M citric acid.

(4) Staining mixture
The staining mixture is the same as above (Section E, 22, c, (3))
(5) Procedure

The gel is prepared and set up in the usual way. Electrophoresis is carried out for approximately 16 hours at about 8 V per linear centimeter of gel with about 10 mA per gel. After completion of electrophoresis, the gel is sliced and stained in a manner similar to that described above (Section E, 32, d).

g. Comment

As with alkaline phosphatase, the acid phosphatase method is nonspecific in the sense that the substrate is synthetic and may have little to do with the physiological function of the enzyme.

Elevated quantitative levels of serum acid phosphatase have long been a useful test for diagnosis of carcinoma of the prostate. The three acid phosphatases in normal serum seem to arise primarily from the prostate (Sur et al., 1962).

Human erythrocytes show an autosomally determined genetic polymorphism in electrophoretic migration (Hopkinson et al., 1964). The electrophoretic variation is also accompanied by quantitative variation.

33. ATPase

a. Reference

This is an unpublished method developed in our laboratory.

b. Basis for the Staining Procedure

The sites of ATPase isozymes are stained according to the following reaction:

$$ATP \xrightarrow{\text{ATPase}} ADP + \text{phosphate}$$

Liberated phosphate is then stained by utilization of the phosphorous method of Fiske and Subbarow (1925). Most ATPase enzymes require Mg^{++}, and some require Na^+ and K^+ for activity.

c. Reagents

(1) Gel buffer
 0.01 M tris
 0.01 M maleic acid
 0.001 M EDTA
 0.001 M $MgCl_2$

The pH is adjusted to 7.6 with 4 N NaOH. Some of the reagents (notably EDTA) will not go into solution until the NaOH is added.

(2) Bridge buffer
 0.1 M tris
 0.1 M maleic acid
 0.01 M EDTA
 0.01 M $MgCl_2$

The pH is adjusted to 7.6 with 4 N NaOH. Note that the gel buffer is a 1:10 dilution of the bridge buffer.

(3) Staining mixture
 0.1 M NaCl
 0.02 M KCl
 0.006 M $MgCl_2$
 0.0015 M ATP
 0.1 M tris, pH 7.6
 0.75 gm Ionagar/100 ml of staining mixture

The Ionagar, NaCl, $MgCl_2$, and KCl are dissolved in ½ to ¾ of the tris buffer and heated to boiling. This mixture is then allowed to cool to 45°C. While it is cooling, the ATP is added to the remainder of the tris buffer. The two mixtures are then added together and stirred. This solution is now ready to pour as an agar overlay. Incubate the gel with the agar overlay for 2–3 hours. After this time remove the agar overlay and apply a second agar overlay.

The second agar overlay is prepared by mixing (per 100 ml of overlay) 10 mg of p-methylaminophenol sulfate and 30 mg of sodium bisulfite in 25 ml of water. Then 250 mg of ammonium molybdate is dissolved in 25 ml of 1.2 N sulfuric acid. 0.75 gm of Ionagar is placed in 50 ml of water and dissolved by boiling. After the latter solution cools to 45°C, the reagents are all mixed, and poured over the gel as an agar overlay.

The final concentration of the reactants in this second overlay is as follows:

0.00029 M p-methylaminophenol sulfate
0.0029 M $NaHSO_3$
0.002 M $(NH_4)_6 Mo_7O_{24} \cdot 4H_2O$ (ammonium molybdate)
0.3 N H_2SO_4
0.75 gm of Ionagar/100 ml staining solution

d. Procedure

The gel is prepared and set up in the usual way. Electrophoresis is carried out for 18–20 hours at approximately 8–10 V per linear centimeter of gel, and about 15 mA per gel. After completion of electrophoresis the gel is sliced in the usual manner. Then the first agar overlay is poured on the gel and allowed to incubate at 37°C for 2–3 hours. During this time phosphorus is being liberated from ATP by the action of the enzyme. Then the first agar overlay is removed and replaced with the second agar overlay. The appearance of greenish-blue bands marks the site of the enzyme.

e. Organisms Studied

This method has been used for *Drosophila* and on commercially purchased apyrase (Table II).

f. Comment

It should be noted that this method can be conveniently combined with other methods described in this chapter using the same buffer system. These include PGM, achromatic regions, and LDH.

It should also be noted that this basic method can be adapted to other phosphate enzymes which liberate phosphate by substituting the appropriate substrate.

34. Hemoglobin

a. REFERENCES

This method is a modification of Smithies' (1959) method (J. Eaton, personal communication).

b. BASIS FOR THE STAINING PROCEDURE

The hemoglobin is stained with an amido-black protein stain.

c. REAGENTS

(1) Gel buffer
 0.05 M tris
 0.007 M boric acid
 0.002 M EDTA
Adjust the pH to 9.0.
 (2) Bridge buffer
 0.2 M sodium barbital
 0.04 M 5,5-diethyl barbituric acid
Heat to boiling to bring reagents into solution and then adjust the pH to 8.6.
 (3) Staining mixture
To make up 110 ml of staining solution, 0.2 gm of amido black are added to 50 ml of water, 50 ml of absolute methanol, and 10 ml of glacial acetic acid.
 (4) Destaining solution
To make up 110 ml of destaining solution, mix 50 ml of water, 50 ml of absolute methanol, and 10 ml of glacial acetic acid.

d. PROCEDURE

The gel is prepared in the usual way. A modification of the usual hemolysate preparation gives somewhat better results. An undiluted hemolysate is prepared by freezing and thawing packed red cells three times. 0.2 ml of a 2% potassium ferricyanide-2% potassium cyanide solution is mixed with 0.2 ml of hemolysate. This mixture is then diluted to 3.0 ml with water and used for electrophoresis. Electrophoresis is carried out for 3–4 hours at 10–12 V per linear centimeter of gel, and about 17 mA per gel. After completion of electrophoresis the gel is sliced in the usual manner. The staining mixture is poured over the gel for several minutes and then the gel is rinsed with the destaining solution several times until the black bands stand out clearly.

e. ORGANISMS STUDIED

This method has been used for the study of mammalian erythrocytes (Table II).

35. General Protein Staining Method

A protein stain can be applied to any of the electrophoretic procedures described in this chapter. If the isozymes are present in relatively high protein concentration, it is often convenient to stain one slice of the gel for protein, and the other for the activity of the enzyme under study. This is particularly true of purified or partially purified enzymes.

The amido-black stain described in the preceding section (Section E, 34) can be used as presented for staining proteins other than hemoglobin. In addition Nigrosin (water soluble) can be used. 0.4 gm of Nigrosin are dissolved in 50 ml of water, 50 ml of absolute methanol, and 10 ml of glacial acetic acid: The gel should be exposed to this stain for 4–10 minutes, depending upon the concentration of protein to be stained. The gel is then rinsed several times with a destaining solution consisting of 50 ml of water, 50 ml of absolute methanol, and 10 ml of glacial acetic acid. Rinsing is carried out until the bands stand out clearly.

F. Other Methods

1. Glutamic-Oxaloacetic Transaminase (GOT)

This method was reported by Nisselbaum and Bodansky (1965) in a study of rabbit reticulocytes and erythrocytes with starch gel. The staining method was adapted from Schwartz *et al.* (1963).

The gel buffer was 0.005 M succinate tris, pH 7.2. The bridge buffer was 0.1 M phosphate, pH 7.2. The staining mixture was 0.04 M DL-aspartic acid, 0.005 M α-ketoglutaric acid, and containing 50 mg pyridoxal phosphate per 100 ml of 0.034 M phosphate buffer pH 7.0. Just before use, 126 mg of Fast Violet B salt is added.

Electrophoresis was carried out for 18 hours at 4 V per linear centimeter of gel and 9–13 mA per gel, and then the sliced gels were stained with the staining mixture until purple bands appear. The color diffuses fairly rapidly.

2. Amylase

This method was reported by Aw (1966) in a study of human urine, saliva, salivary glands, and pancreatic extracts.

One microliter of material was applied to cellulose acetate strips with veronal as a bridge buffer (pH and strength not given). Electrophoresis was carried out at 200 V (total) at 0.4 mA per centimeter width for 1 hour. For isozyme identification starch plates were prepared by thin layer chromatography apparatus on glass plates. Connaught starch of the type used for starch gel electrophoresis was made up in the concentration recommended on the label in barbital-sodium acetate buffer containing 0.02 M calcium chloride to prevent phosphorolytic digestion of the starch.

After electrophoresis, the cellulose acetate strips were carefully laid on the starch plates and incubated at 37°C for 30 minutes. Then the strips were removed and the glass plates stained with 0.02 N iodine. The amylase isozymes were indicated by a clear area in the purple background of undigested starch.

3. Ceruloplasmin

Starch gel electrophoresis is the principal method employed for the identification and classification of serum ceruloplasmin variants. Either horizontal or vertical starch gel electrophoresis can be used although better definition of the variants is accomplished with the former. The method is a modification (Shreffler et al., 1967; Shokeir, 1969) of that described by Smithies (1955). For optimal resolution, 11.8 gm Electrostarch (Electrostarch Co., Madison, Wisconsin) per 100 ml borate gel buffer (0.012 M boric acid, 0.0074 M sodium hydroxide, pH 9.55) is used. The bridge buffer is 0.21 M boric acid, 0.085 M sodium hydroxide, pH 9.0. Electrophoresis is performed at 4°C for 21 hours at 8 V per linear centimeter of gel. Serum samples are routinely diluted 1:3 with water before electrophoresis to diminish the intensity of ceruloplasmin staining, enhance resolution, and provide better distinction of the variants.

Ceruloplasmin bands are stained by a modification of the method of Owen and Smith (1961). Gels are sliced in half horizontally upon completion of electrophoresis. The bottom layer is stained by incubation in a solution of 0.1% o-dianisidine and 30% ethanol in 0.04 M acetate buffer, pH 5.5, for 1 to 1½ hours at 37°C. At this stage, ceruloplasmin bands are well defined when examined on an X-ray type illuminator. The gels can be photographed at this stage. Polymorphic variation has been observed in the American Negro, but only rare variants occur in the American Caucasian population (Shreffler et al., 1967).

Distinction of different ceruloplasmin variants has been substantially enhanced by studying response to inhibition of their quantitatively determined oxidase activities by sodium cyanide and sodium azide (Shokeir, 1969). By the addition of 0.10 ml of 0.004 M sodium cyanide to 0.15 ml serum of the common phenotype (CpB) 69% reduction of oxidative activity could be produced, while the same level of the inhibitor produced only 39% inhibition of the variant phenotype CpA. The heterozygous type $CpAB$ displayed 49% inhibition under these conditions. These percentages of inhibition in the three phenotypic classes were consistently obtained and indicate contribution of the Cp^A and Cp^B alleles to the ceruloplasmin protein of the $CpAB$ heterozygote in a ratio of 2:1 (Shokeir, 1969). Corresponding results were obtained by using sodium azide inhibition, though the differential response of the variants is not as pronounced as in the case of cyanide inhibiton. The addition of 0.05 ml of 0.01% sodium azide to 0.15 ml serum of the common phenotype CpB is accompanied by 50% inhibition of oxidase activity, whereas the same level of inhibitor brought about only 44% reduction of activity of the variant type CpA. The heterozygous type $CpAB$ sustained 46% inhibition, again pointing to a make up of ceruloplasmin protein in $CpAb$ heterozygote of two parts CpA protein and one part CpB protein. This conclusion was further corroborated when a mixture of two parts serum for a CpA individual and one part serum from a CpB individual exhibited an inhibition response to both sodium cyanide and sodium azide very similar to that of $CpAB$ serum.

REFERENCES

Aebi, H., and Suter, H. (1969). In "Biochemical Methods in Red Cell Genetics" (J. Yunis, ed.). Academic Press, New York, pp. 255–285.
Allen, J. M. (1963). *J. Histochem. Cytochem.* **11**, 542.
Allen, J. M., and Hynick, G. (1963). *J. Histochem. Cytochem.* **11**, 169.
Anstall, H. B., Lapp, C., and Trujillo, J. M. (1966). *Science* **154**, 657.
Aw, S. E. (1966). *Nature* **209**, 299.
Baughan, M. A., Valentine, W. N., Paglia, M. D., Ways, P. O., Simon, E. R., and DeMarsh, O. B. (1968). *Blood* **32**, 236.
Beckman, L., and Johnson, F. M. (1964). *Hereditas* **51**, 221.
Beckman, L., Bjorling, G., and Christodoulou, C. (1966a). *Acta Genet. Statist. Med.* **16**, 223.
Beckman, L., Bjorling, G., and Christodoulou, C. (1966b). *Acta Genet. Statist. Med.* **16**, 122.

Bell, J. L., and Baron, D. N. (1962). *Biochem. J.* **82**, 5P.
Bernsohn, J., Barron, K. D., and Hess, A. (1961). *Proc. Soc. Exptl. Biol. Med.* **108**, 71.
Bourne, J. G., Collier, H. O. J., and Somers, G. F. (1952). *Lancet* i, 1229.
Bowman, J. E., Frischer, H., Ajmar, F., Carson, P. E., and Gower, M. K. (1967). *Nature* **214**, 1156.
Boyer, S. H. (1961). *Science* **134**, 1002.
Brewer, G. J. (1967). *Am. J. Human Genet.* **19**, 674.
Brewer, G. J. (1969). *In* "Biochemical Methods in Red Cell Genetics" (J. Yunis, ed.). Academic Press, New York, pp. 139–163.
Brewer, G. J., and Knutsen, C. A. (1968). *Science* **159**, 650.
Brewer, G. J., and Sing, C. F. (1969). *In* "Biochemical Methods in Red Cell Genetics" (J. Yunis, ed.). Academic Press, New York, pp. 377–390.
Brewer, G. J., Gall, J. C., Honeyman, M. S., Gershowitz, H., Shreffler, D. C., Dern, R. J., and Hames, C. (1967a). *Biochem. Genet.* **1**, 41.
Brewer, G. J., Bowbeer, D. R., and Tashian, R. E. (1967b). *Acta Genet. Statist. Med.* **17**, 97.
Brewer, G. J., Eaton, J. W., Knutsen, C. A., and Beck, C. C. (1967c). *Biochem. Biophys. Res. Commun.* **29**, 198.
Brewer, G. J., Sing, C. F., and Sears, E. R. (1969). *Proc. Natl. Acad. U. S.* (in press).
Bucher, T., and Klingenburg, M. (1958). *Angew. Chem.* **70**, 552.
Cahn, R. D., Kaplan, N. O., Levine, L., and Zwilling, E. (1962). *Science* **136**, 962.
Campbell, D. M., and Maas, D. W. (1962). *Proc. Assoc. Clin. Biochem.* **2**, 10.
Davidson, R. G., and Cortner, J. A. (1967a). *Nature* **215**, 761.
Davidson, R. G., and Cortner, J. A. (1967b). *Science* **157**, 1569.
Davidson, R. G., Nitowsky, H. M., and Childs, B. (1963). *Proc. Natl. Acad. Sci. U. S.* **50**, 481.
Delbruck, A., Zebe, E., and Bucher, T. (1959). *Biochem. Z.* **331**, 273.
Detter, J. C., Ways, P. O., Giblett, E. R., Baughan, M. A., Hopkinson, D. A., Povey, S., and Harris, H. (1968). *Ann. Human Genet.* **31**, 329.
Eaton, G. M., Brewer, G. J., and Tashian, R. E. (1966). *Nature* **212**, 944.
Eppenberger, M. E., Eppenberger, H. M., and Kaplan, N. O. (1967). *Nature* **214**, 239.
Evans, F. T., Gray, P. W. S., Lehmann, H., and Silk, E. (1952). *Lancet* i, 1225.
Fildes, R. A., and Harris, H. (1966). *Nature* **209**, 5020.
Fiske, C. H., and Subbarow, Y. (1925). *J. Biol. Chem.* **66**, 375.
Goldberg, E. (1963). *Science* **139**, 602.
Gomori, G. (1952). "Microscopic Histochemistry," p. 211. Univ. of Chicago Press, Chicago, Illinois.
Grell, E. H., Jacobson, K. B., and Murphy, J. B. (1965). *Science* **149**, 80.
Henderson, N. S. (1965). *J. Exptl. Zool.* **158**, 263.
Hirsch, C. A., Rasminsky, M., Davis, B. D., and Lin, E. C. C. (1963). *J. Biol. Chem.* **238**, 3770.
Holmes, E. W., Jr., Malone, J. L., Winegrad, A. J., and Oski, F. A. (1967). *Science* **156**, 646.
Hopkinson, D. A., and Harris, H. (1965). *Nature* **208**, 410.
Hopkinson, D. A., Spencer, N., and Harris, H. (1964). *Am. J. Human Genet.* **16**, 141.
Kalow, W., and Genest, K. (1957). *Can. J. Biochem. Physiol.* **35**, 339.
Kaplan, J. C. (1968). *Nature* **217**, 256.

Chapter 5. Specific Electrophoretic Systems

Kaplan, J. C., and Beutler, E. (1968). *Science* **159**, 215.
Kaplan, N. O. (1961). *In* "Mechanism of Action of Steroid Hormones" (C. A. Villee and A. A. Engel, eds.). Macmillan (Pergamon), New York, pp. 247–255.
Kaplan, N. O. (1963). *Bacteriol. Rev.* **27**, 155.
Kaplan, N. O., and Ciotti, M. M. (1961). *Ann. N. Y. Acad. Sci.* **94**, 701.
Kitto, G. B., Kottke, M. E., Bertland, L. H., Murphy, W. H., and Kaplan, N. O. (1967). *Arch. Biochem. Biophys.* **121**, 224.
Knutsen, C. A., Sing, C. F., and Brewer, G. J. (1969). *Biochem. Genet.* **3**, 475.
Koen, A. L., and Shaw, C. R. (1965). *Biochem. Biophys. Res. Commun.* **15**, 92.
Koler, R. D., Bigley, R. H., Jones, R. T., Rigas, D. A., Vanbellinghen, P., and Thompson, P. (1964). *Cold Spring Harbor Symp. Quant. Biol.* **29**, 213.
Kowlessar, O. D., Haeffner, L. J., and Sleisenger, M. H. (1960). *J. Clin. Invest.* **39**, 671.
Latner, A. L. (1965). *In* "Enzymes in Clinical Chemistry" (R. Ruyssen and L. Vandenriessche, eds.), p. 110. Elsevier, Amsterdam.
Latner, A. L., and Skillen, A. W. (1962). *Proc. Assoc. Clin. Biochem.* **2**, 3.
Latner, A. L., and Skillen, A. W. (1968). "Isoenzymes in Biology and Medicine." Academic Press, New York.
Lawrence, S. H., Melnick, P. J., and Weimer, H. E. (1960). *Proc. Soc. Exptl. Biol. Med.* **105**, 572.
Lehmann, H., and Liddell, J. (1964). *Progr. Med. Genet.* **3**, 75.
Long, W. K. (1966). *Proc. 3rd Intern. Congr. Human Genet., Chicago* p. 59. (Abstr.)
Long, W. K. (1967). *Science* **155**, 712.
Lowenstein, J. M., and Smith, S. R. (1962). *Biochim. Biophys. Acta* **56**, 385.
Luck, D. J. L., and Reich, E. (1964). *Proc. Natl. Acad. Sci. U. S.* **52**, 931.
Markert, C. L., and Hunter, R. L. (1959). *J. Histochem. Cytochem.* **7**, 42.
Markert, C. L., and Moller, F. (1959). *Proc. Natl. Acad. Sci. U. S.* **45**, 753.
Monis, B. (1964). *J. Histochem. Cytochem.* **12**, 869.
Monis, B. (1965). *Federation Proc.* **24**, 681.
Motulsky, A. G. (1960). *Human Biol.* **32**, 28.
Munkres, K. D. (1965). *Arch Biochem. Biophys.* **112**, 347.
Nishimura, E. T., Carson, N., and Kobara, T. Y. (1964). *Arch. Biochem. Biophys.* **108**, 452.
Nisselbaum, J. S., and Bodansky, O. (1965). *Science* **149**, 195.
Ohno, S., Payne, H. W., Morrison, M., and Beutler, E. (1966). *Science* **153**, 1015.
Owen, J. A., and Smith, H. (1961). *Clin. Chim. Acta* **6**, 441.
Pineda, E. P., Goldberg, J. A., Banks, B. M., and Rutenberg, A. M. (1960). *Gastroenterology* **38**, 698.
Reich, E., and Luck, D. J. L. (1966). *Proc. Natl. Acad. Sci. U. S.* **55**, 1600.
Ressler, N., and Stitzer, K. (1967). *Biochim. Biophys. Acta* **146**, 1.
Robinson, J. C., Pierce, J. E., and Goldstein, D. P. (1965). *Science* **150**, 58.
Rosalki, S. B. (1965). *Nature* **207**, 414.
Rossi, C., Hauber, J., and Singer, T. P. (1964). *Nature* **204**, 167.
Sactor, B. (1959). *Proc. 4th Intern. Congr. Biochem., Vienna, 1958* **12**, 138.
Sandler, M., and Bourne, G. H. (1961). *Exptl. Cell Res.* **24**, 174.
Sandler, M., and Bourne, G. H. (1962). *Nature* **194**, 389.
Scandalios, J. G. (1967). *J. Heredity* **58**, 153.
Scandalios, J. G. (1969). *Biochem. Genet.* **3**, 37.
Scott, E. M., and Griffith, I. V. (1959). *Biochim. Biophys. Acta* **34**, 584.

Schwartz, M. K., Nisselbaum, J. S., and Bodansky, O. (1963). *Am. J. Clin. Pathol.* **40**, 103.
Shannon, L. M. (1968). *Ann. Rev. Plant Physiol.* **19**, 187.
Shaw, C. R. (1966). *Science* **153**, 1013.
Shaw, C. R., and Barto, E. (1963). *Proc. Natl. Acad. Sci. U. S.* **50**, 211.
Shaw, C. R., and Koen, A. L. (1965). *J. Histochem. Cytochem.* **13**, 431.
Shokeir, M. (1969). Ph.D. Thesis, Univ. of Michigan, Ann Arbor, Michigan.
Shows, T. B., and Ruddle, F. H. (1968). *Isozyme Bull.* **1**, 25.
Shows, T. B., Tashian, R. E., Brewer, G. J., and Dern, R. J. (1964). *Science* **145**, 1056.
Shreffler, D. C., Brewer, G. J., Gall, J. C., and Honeyman, M. S. (1967). *Biochem. Genet.* **1**, 101.
Smith, E. E., and Rutenberg, A. M. (1963). *Nature* **197**, 800.
Smith, E. E., and Rutenberg, A. M. (1966). *Science* **152**, 1256.
Smithies, O. (1955). *Biochem. J.* **61**, 629.
Smithies, O. (1959). *J. Biol. Chem.* **71**, 585.
Spencer, N., Hopkinson, D., and Harris, H. (1964). *Nature* **204**, 742.
Standardization of Procedures for the Study of Glucose-6-Phosphate Dehydrogenase (1967). *World Health Organ. Tech. Rept. Ser.* **366**.
Sur, B. K., Moss, D. W., and King, E. J. (1962). *Proc. Assoc. Clin. Biochem.* **2**, 11.
Tanaka, T., Harano, Y., Morimura, H., and Mori, R. (1965). *Biochem. Biophys. Res. Commun.* **16**, 319.
Tashian, R. E. (1969). In "Biochemical Methods in Red Cell Genetics" (J. Yunis, ed.), Academic Press, New York, pp. 307–334.
Tashian, R. E., and Shaw, M. W. (1962). *Am. J. Human Genet.* **14**, 295.
Tashian, R. E., Riggs, S. K., and Yu, Y. S. L. (1966). *Arch. Biochem. Biophys.* **117**, 320.
Thorne, C. J. A. (1960). *Biochim. Biophys. Acta* **42**, 175.
Thorup, O. A., Jr., Strole, W. B., and Leavell, B. S. (1961). *J. Lab. Clin. Med.* **58**, 122.
Tsao, M. U. (1960). *Arch. Biochem. Biophys.* **90**, 234.
Ursprung, H., and Leone, J. (1965). *J. Exptl. Zool.* **160**, 147.
Van derHelm, H. J. (1962). *Nature* **194**, 773.
Vesell, E. S., and Bearn, A. G. (1958). *J. Clin. Invest.* **37**, 672.
Wieland, T., Pfleiderer, G., Haupt, I., and Worner, W. (1959). *Biochem. Z.* **331**, 103.
Wilkinson, J. H. (1965). "Isoenzymes." Lippincott, Philadelphia, Pennsylvania.
Williams, R. A. D. (1964). *Nature* **203**, 1070.
Yakulis, V. J., Gibson, C. W., and Heller, P. (1962). *Am. J. Clin. Pathol.* **38**, 378.
Yonetani, T. (1963). In "The Enzymes" (P. D. Boyer, H. Lardy, and K. Myrbäck, eds.), Academic Press, New York, pp. 41–70.

Chapter 6

Present Applications and the Future of Isozymology

A. Introduction

In this chapter those areas which in the opinion of the author are, or may become, the "cutting edges" of the science of isozymology will be pointed out. With respect to future developments the areas emphasized will, of course, be somewhat speculative, and perhaps biased to some extent by the interests of the author. Nevertheless, it should give the reader an indication of the breadth and depth of this field and the scope of the problems to which the techniques described in this book can be applied. Hopefully it will also suggest to some investigators profitable future courses of research.

B. Clinical Applications

1. Current Clinical Applications

a. Serum Isozymes

The most notable example of the application of isozyme techniques to problems in clinical medicine involves serum lactic dehydrogenase (LDH) (Chapter 5, Section E, 8). The application of studies of serum LDH isozymes to clinical medicine has been reviewed by Latner and Skillen (1968) and by Wilkinson (1965). Human LDH patterns usually consists of a series of five isozymes with LDH-1 being the most anodal. The subunit composition of LDH has been well defined. The molecule is a tetramer of two randomly combining subunits called A and B. LDH-1 consists of four B subunits, LDH-2 of three B and one A subunits, LDH-3 of two B and two A subunits, LDH-4 of one B and three A subunits, and LDH-5 of four A subunits.

Tissues vary in the relative amounts of the A and B subunits synthesized, leading to considerable tissue variation in LDH isozyme patterns. It is this fact which is the basis for clinical application. If a tissue is damaged and cellular injury occurs, such as with a coronoary occlusion producing myocardial infarction, some of the intracellular enzymes, including LDH, will escape into the serum. At this point, an LDH isozyme pattern of the serum will reveal the isozymes of that particular tissue. However, many tissues have similar patterns, so this method by itself is not diagnostic. For example, both heart tissue and erythrocytes have predominantly LDH-1. In the absence of other information, then, one cannot say that the presence of an LDH-1 pattern in the serum is due to either hemolysis, on the one hand, or coronary occlusion, on the other (see Section B, 2, this chapter, for a further expansion of this area with respect to future applications).

The most important clinical use for the LDH isozymes has been as an indicator of heart or liver disease. As pointed out above, LDH-1 predominates over LDH-5 in heart muscle. In the presence of myocardial damage, the serum level of LDH-1 and, to a certain extent, LDH-2, goes up in a few hours. This pattern fades after a few days as healing takes place, but it reappears if a second infarction occurs. The type of serum pattern following myocardial infarction is not specific, even for the type of cardiac damage. For example, it also occurs in acute rheumatic myocarditis. It will also occur, of course, with hemolytic anemia, in which erythrocytes are being destroyed and are releasing LDH-1 into the serum. In megaloblastic anemia the LDH-1 band is also very pronounced in the serum. However, the appearance of a myocardial pattern of serum LDH isozymes in the presence of other clinical indications of a possible infarction is strong evidence that myocardial damage has taken place. This technique has proven quite useful for both diagnosis and in following the resolution of the damage.

In liver disease LDH-5 predominates in the serum. Since LDH-1 predominated in erythrocytes, the evaluation of LDH serum isozymes is of considerable use in differentiating hyperbilirubinemia of hemolytic and hepatic origins. Although LDH-5 is elevated in jaundice of all types caused by diseases of the liver, it is by far the most prominent in hepatocellular disease, such as that due to viral hepatitis.

LDH isozyme patterns have also been evaluated in a number of other diseases (reviewed by Latner and Skillen, 1968). An interesting possible application is the use of the serum LDH patterns to detect the early stages of homograft rejection. Patients who have subsequently rejected their kidney homografts are reported to have an increase in serum LDH-1 and LDH-2 prior to the rejection (Prout et al., 1964).

A certain amount of clinical work has also been done with other serum isozyme systems (reviewed by Latner and Skillen, 1968), including work with alkaline phosphatase, aspartate aminotransferase, creatine phosphotransferase, acid phosphatase, leucine aminopeptidase, and esterase. However, for the most part, the observations on these systems have been more fragmentary and have not had the impact of the LDH studies.

An important aspect of the clinical application if isozyme studies is the addition of characterization of the isozymes beyond electrophoretic migration. This has been done rather extensively with the LDH system, primarily to simplify the detection of specific isozymes. For example, heat stability (65°C and 57°C for 30 minutes; Wroblewski and Gregory, 1961), oxalate and urea inhibition (Emerson and Wilkinson, 1965), activity with oxobutyrate as a substrate, and solvent precipitation techniques all have been suggested for this purpose. In other words, such tests can be substituted for the electrophoretic procedure, since they differentially affect the isozymes, and the remaining activity is a reflection of the type of isozyme present. However, beyond simplification of isozyme detection, isozyme characterization may also be useful in differentiating isozymes from different tissues which migrate to the same or nearly the same place on the gel (see Section B, 2, this chapter).

b. Neoplastic Disease

Variation in isozyme patterns have been observed in neoplasia. In general malignant tumors (solid type) have predominantly slower moving LDH isozymes (Latner and Skillen, 1968). This also applies to effusions from malignant tumors (Richterich et al., 1962). Often serum isozyme patterns are a reflection of the tumor type and can be used to evaluate the status of the disease and the effect of therapy (Starkweather et al., 1966a).

The reverse type of patterns (to solid tumors) tend to be the case in leukemia. The faster moving or intermediate LDH isozymes predominate in bone marrow cells in acute granulocytic and acute lymphocytic leukemias, and the slower isozymes are most marked in chronic leukemias of both types (Starkweather et al., 1966b). Patterns in bone marrow samples return toward normal with treatment (H. H. Spencer and W. H. Starkweather, personal communication). As the disease relapses, the abnormal pattern begins to reappear. It is the feeling of this group that evaluation of LDH isozyme patterns is a useful means of following the disease, allowing the physician to predict relapse and remission some time before morphological evidence heralds such events.

Investigations in our laboratory have been directed toward leukemic white cells. We have observed differences in hexokinase patterns, with leukocytes from chronic granulocytic leukemia having several more bands than cells from chronic lymphocytic leukeima (Eaton et al., 1966; Eaton and Brewer, unpublished observations). There are also rather marked differences in esterase patterns in peripheral leukocytes of lymphocytic and granulocytic leukemias (C. Coleman and G. J. Brewer, unpublished observations). As experience develops, it seems likely that isozyme patterns will be of considerable assistance in diagnosis of the subtypes of leukemia which are often hard to separate on morphological grounds, particularly in the acute phases. Also, it seems likely that following the abnormal isozyme patterns will be of value in following the course of the disease and its response to treatment. (See Section B, 2 of this chapter for a discussion of future applications.)

A unique use of isozymes has been in the study of a basic and very important question regarding the cellular etiology of malignancy. The question is whether or not tumors arise from a single cell or from a tissue area involving many cells. Linder and Gartler (1965) have made use of the glucose-6-phosphate dehydrogenase (G-6-PD) electrophoretic system and the single active-X hypothesis (see Chapter 1, Section B for an explanation of the latter) to study this question. G-6-PD is controlled by an X-linked gene in the human, and an electrophoretic polymorphism exists in the Negro. By screening of Negro females with tumors to detect individuals heterozygous for G-6-PD alleles, it is possible to carry out very significant experiments on the tumor after it is removed or biopsied. If the tumor is derived from one cell, it should show a single electrophoretic band, while if it is derived from several cells, it should show both bands (cells reproduce true to type with respect to X-inactivation during somatic growth). Linder and Gartler (1965) have shown that tissues from uterine leiomyomas of Negro females heterozygous for G-6-PD electrophoretic alleles show single bands of G-6-PD, indicating their origin from a single cell. On the other hand, in hereditary multiple trichoepithelioma, an autosomal dominant condition characterized by multiple tumors of the skin, studies of G-6-PD deficiency alleles revealed the orgin of the tumors from multiple cells (Gartler et al., 1966). This important type of investigation needs to be extended to other neoplastic conditions such as leukemia and lymphomas.

2. Future Clinical Applications

The potential benefit from the utilization of isozyme techniques in clinical medicine is great. It is becoming increasingly apparent that isozyme diversity from one tissue to another is very common. This means that it should be eventually possible, by the use of enough isozyme systems, to uniquely characterize every tissue. This will be done with fewer isozyme systems, no doubt, if a few enzyme characterization steps, such as substrate specificity and inhibition studies, are also utilized. This implies, for example, that if a clinical laboratory with some 20 well-developed isozyme systems, including a few enzyme characterization procedures for some of the systems, were to be given an extract of tissue, it would be possible to make positive identification of that tissue from isozyme patterns. If this were true, the same capabilities would also eventually pertain to the detection of almost any damaged organ by study of serum isozymes. There are, of course, a few additional difficulties in this case compared to the identification of a tissue from its extract. These include the release of a sufficient quantity of the isozymes into the serum from the damaged organ and the requiremnt that the isozymes be sufficiently stable to persist in the serum in high enough concentration to detect. However, the general principle still holds. By utilization of sufficient isozyme systems, particularly if sensitivity of the procedures is maximized, injured tissue of any type should be capable of identification.

It seems to the author, however, that the potential of isozymology in clinical medicine is not being developed at the rate at which it could be developed. The examples cited earlier show how the use of isozyme techniques can be useful in a number of areas of medicine, but they consist, for the most part, of applications of a single technique. There is a strong tendency of many investigators and clinical laboratories to be single-system oriented. We strongly urge that increasing numbers of laboratories capable of carrying out isozyme techniques develop a multisystem approach, because with the possible exception of the LDH system, the usefulness of any single system is likely to be rather limited. As pointed out earlier, many tissues have overlapping isozyme patterns, and it is freqeuntly impossible to differentiate the source of one set of serum isozymes from another when a single system is being used. It is our belief that by the use of a multisystem approach it will be possible, as suggested above, to "fingerprint" each tissue for its isozyme characteristics. It should be emphasized that the strength of this approach will be increased greatly if, in addition to simple electrophoretic analysis, a few simple enzymes characterization steps are carried out. It is obvious that as the complexity of studies increases, it is important, in an attempt to maximize efficiency, to pay heed to the suggestions in Chapter 7 for scoring, data storing, data retrieval, and data evaluation.

One of the problems hindering the development of a multisystem approach is the lack of sufficient overlap and communication between the isozymologist, the clinical pathologist, and the clinician. Those individuals with widespread experience in the field of isozymology tend to be interested in basic problems and not in clinical applications. Clinical pathologists and clinicians have not tended to pick up a large number of isozyme techniques rapidly and to apply them to clinical problems. Obviously there is a need for the development of a class of clinical investigator interested in using isozyme techniques and exploring, in depth, a multisystem approach.

In a sense, nature has given the physician, in the isozymic diversity of tissues, an important handle to assist him in the diagnosis and care of specific tissue diseases. It is up to the isozymologist, clinical pathologist, and the physicians themselves to see that a multisystem approach utilizing as many new methods as possible gains early widespread application to clinical problems.

C. Somatic Cell Genetics

Isozyme techniques are becoming increasingly useful as a tool in the important field of somatic cell (tissue culture) genetics. Perhaps the most dramatic example of this use was the one cited in Chapter 1, Section B, concerning the demonstration by Davidson *et al.* (1963), that the single active-X (Lyon hypothesis) was applicable to the G-6-PD locus of human females.

The techniques of cell fusion and cell hybridization in tissue culture complemented with isozyme studies, would seem to offer much promise for the study of many important cellular and genetic questions. It has been found that cultures treated with inactivated Sendai virus undergo cell fusion (cytoplasmic but not nuclear fusion) at a much increased frequency. By fusing cells with different electrophoretic migration of specific enzymes, including cells from different interspecific sources, the evaluation of the formation of heteropolymeric molecules will give answers to such questions as whether or not there is, in general, heteropolymer formation when two types of subunits are being synthesized in the same cytoplasm of fused cells. This is an approach to the question of whether cytoplasmic integration actually takes place in fused cells. If it does take place in closely related species, does it also take place in cells from widely diverse species? How widely divergent must the species be before heteropolymer formation fails to occur? And can this approach be used to construct phylogenetic relationships?

144 / Chapter 6. Present and Future of Isozymology

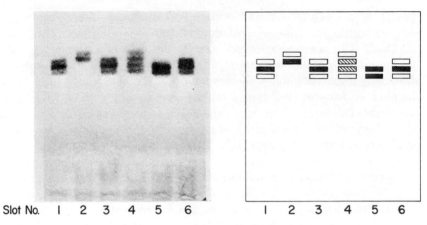

Figure 11. On the left is a photograph of a G-6-PD stained starch gel demonstrating the first human-human hybridization. On the right is an explanatory diagram. Slot 2 contains material from line D98/AH (G-6-PD type A) and slot 5 contains material from line AUC(G-6-PD type B). Slot 4 contains a mixture of material from the two cell lines. Slot 1, 3, and 6 contain material from a cloned cell line indicating hybridization. Note the probable formation of a relatively intense heteropolymer molecule in the hybrid line. This does not ordinarily occur *in vivo* because only one X-chromosome is active in each female cell. Photographic portion reproduced with permission of Selagi et al. (1969).

Cell hybridization indicates nuclear fusion as well as cytoplasmic fusion. While this is much rarer than cell fusion, it does happen spontaneously on occasion and in a small fraction of cells fused by means of treatment with inactivated Sendai virus. Frequently, chromosomes are eliminated from the hybrid cell line until, after a time, a fairly stable line may be derived consisting a mixture of chromosomes from the two parental cells. In crosses between established mouse lines and diploid human lines, most of the mouse chromosomes are retained while only one or two chromosomes persist from the human parent (Migeon and Miller, 1968; Weiss and Green, 1967). Persisting human markers, such as isozyme markers, can then be attributed to the remaining chromosome(s), and in this way linkage groups can be gradually identified. This approach may solve the heretofore very difficult problem of studying linkage of human genes. Of course heteropolymeric molecules from hybrid cells can also be studied and the results compared with results obtained from fused cells. Figure 11 illustrates the use of G-6-PD isozyme markers to characterize the first human–human hybrid cell line (Selagi et al., 1969). Note the probable formation of a heteropolymer molecules in the hybrid cell line. This does not usually occur *in vivo* because ony one X-chromosome is active in each female cell.

In addition to the more or less pragmatic applications of the use of isozyme techniques in somatic cell genetics described in this section, the use of these methods for following the production of specific gene products offers much promise for one of the major theoretical areas of somatic cell genetics. It is hoped that eventually these diploid cell cultures can be manipulated in as many ways and with as much ease as the microbial geneticist now manipulates bacterial cultures. The isozyme markers will provide a convenient method to study these cultures and begin to lay down rules of the methods of regulation of gene expression in diploid cells. Of course, if the regulation of gene expression in diploid cells can be understood, a host of practical applications would follow.

D. Tissue–Organ and Intracellular Differentiation

One of the major features of isozyme work so far is the occurrence of diverse isozymes in different tissues of complex organisms. This has already been discussed in some detail in relation to the use of serum isozymes in clinical applications (Section B, this chapter). It will also be discussed in relation to evolution (Section I, this chapter). An illustration of such isozymic diversity among tissues of *Drosophila* is shown in Figure 12, part A (Knutsen *et al.*, 1969) which concerns hexokinase isozymes. In general, it is felt by most workers that tissue isozymic diversity subserves metabolic roles that differ slightly from one tissue to another. It is also possible that many isozymes are compartmentalized within cells, that is one isozyme for the mitochondria, one for the cell wall, and so on. In this situation the catalytic functions might not differ very much between isozymes, but charge differences in other parts of the molecule might be involved in intracellular organization. It might be quite feasible for evolution to build up a series of isozymes by means of gene duplication which then become adapted to different compartments of the cell, or to different organs, by virtue of mutation in areas of the molecule other than the active site. That molecular charge may be more than a random event (of benefit to the isozymologist!) is suggested by the rather specific charge characteristics of a number of enzymes across many species. A good example is alcohol dehydrogenase of mammalian tissues, which always migrates cathodally. Carbonic anhydrases I and II, which can be identified by means other than migration, always have the same relative position to one another after electrophoresis. Thus charge characteristics may be an important facet of intracellular organization. And, of course, the charge characteristics of the molecule may not depend strictly upon the primary structure, but perhaps can be modified by the organism by the attachment of small charged molecules such as sialic acid.

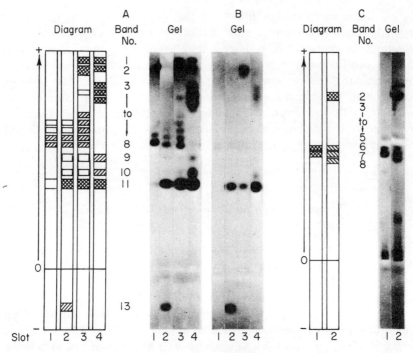

Figure 12. Photograph of starch gel hexokinase patterns of *Drosophila robusta*. Part A (diagram on the left, photograph on the right): slot 1, ovary; slot 2, testes; slot 3, muscle; slot 4, digestive tissue. Part B: identical gel to Part A stained with fructose as substrate rather than glucose. Part C (diagram on the left, photograph on the right): slot 1, larvae; slot 2, pupa. Reproduced with permission of Knutsen et al. (1969).

E. Developmental Genetics

Considerable ontogenetic variation in isozyme patterns has been reported. In fact, probably more than half of the isozyme systems in the typical organism will show some variation in patterns during the life cycle. The greatest variation ordinarily occurs early, coinciding, in general, with the time of the greatest ontogenetic morphological variation. For example, most of the variation in *Drosophila* hexokinase isozymes appear in the larval to pupal stages (Figure 12, part C) (Knutsen et al., 1969). In mammalian systems the same general rule applies. For example, in the human there is not much ontogenetic isozyme change after birth, but a considerable amount will probably be observed when early fetal specimens are investigated in more detail. In the case of human hemoglobins five different types of chains (alpha, beta, delta, gamma, and epsilon) are synthesized in the embryo at different times prior to birth; after birth no new types of chains are synthesized. It will be of interest to see the results of combined chromosome and isozyme studies in fetuses aborted early in pregnancy. A high incidence of chromosomal abnormalities occurs in such material, and it may be possible to locate structural genes for isozymes on specific choromsomes by concurrently studying isozyme patterns.

Isozyme markers should continue to provide a convenient tool with which to advance our understanding of developmental genetics. It will be of interest for example, to look for developmental covariation of isozymes which are related in metabolic pathways (see also Section F, 2, this chapter, for a discussion of covariation of isozymes). If two metabolically related isozyme systems should follow developmental sequences which are in some way synchronized, this may provide a useful experimental tool to study genetic regulation and differentiation in higher organisms. Observations such as this will allow us to begin to put in perspective the biological significance of isozymes.

F. Genetic Variation

1. The Use of Genetic Variation in the Study of Subunit Composition and Structural Relationships of Isozymes

There are two reciprocal relationships in the discussion of the joint topics of isozyme techniques and genetic variation. The first is the use of genetic variation to analyze the subunit composition and structural relationships of isozymes, and the second is the use of isozyme techniques in the study of genetic variation of populations. We will consider the first topic in this section. This area has been admirably reviewed by Shaw (1964, 1965) who also discovered the first lactic dehydrogenase genetic variant (Shaw and Barto, 1963) and used the patterns to confirm the subunit hypothesis previously indicated (Markert, 1962, 1963; Cahn *et al.*, 1962) on the basis of subunit dissociation and reassociation studies.

First, it should be recognized that the isozyme pattern itself, observed prior to the use of genetic variation, can suggest likely types of subunit composition and polymeric structure of isozymes. For example, many enzymes exist as dimers. If two types of subunits (with differing net charge) are produced in each organism a 2-band (constazyme-type) pattern may be present if the two types of subunits will not randomly

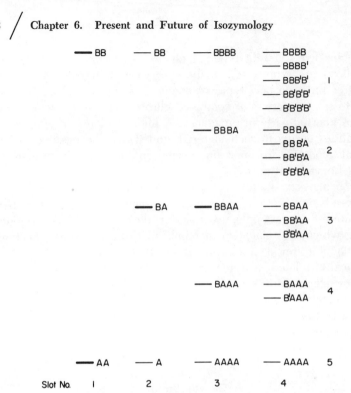

Figure 13. Diagram of possible combinations of isozyme bands according to subunit composition. Slot 1, nonrandom dimeric combinations of two types of subunits. Slot 2, random dimeric combinations of two types of subunits; note the heteropolymer formation. Slot 3, random tetrameric combinations of two types of subunits, such as found in the LDH system. Slot 4, formation of 15 bands when a mutation is superimposed on the situation in slot 3, such that the organism is heterozygous for one of the types of subunits.

combine (Figure 13, slot 1). If the two types of subunits will randomly combine, a 3-band constazyme pattern may be present, with the heteropolymeric molecule showing an intermediate migration (Figure 13, slot 2). Typically, if the subunits are produced in approximately equal amounts, and if their specific activities are approximately equal, the ratio of activity in the 3 bands will be 1:2:1, reflecting the proportion of each type isozyme which will form by chance.

Similar considerations apply to more complex subunit structures, such as the mammalian lactic dehydrogenase (LDH) system. LDH consists of a tetramer of two types of randomly combining subunits. A 5-band pattern is consistently seen, explained by the subunit composition of isozymes shown in Figure 13, slot 3). Shaw (1964; and personal communication from C. Markert to Shaw) has presented a table and formula for predicting the number of isozymes to be observed with varying numbers of types of randomly combining subunits and varying numbers of subunits per polymeric molecule (Figure 14). Assuming equal specific activity of all types of polymers, the relative strengths of bands derived from two types of randomly combining subunits will be a simple expansion of the binomial. In the dimer, this is 1:2:1, in the tetramer 1:4:6:4:1, etc.

Although the number of bands and the relative activities of the bands may give useful early clues to the subunit composition of a series of isozymes, other evidence is required before anything reasonably definitive can be concluded. Alternate explanations for isozymes patterns are always possible. One approach is enzyme purification and study of subunits by subunit dissociation and reassociation techniques. However, these methods are technically difficult and laborious, and often not possible because of insufficient material, instability of the enzyme, or other problems. The finding of a genetic variant, if such are not too rare, is usually a much easier method of obtaining information. Even if the enzyme shows only 1 band of activity ordinarily, a heterozygote, because it now produces two types of subunits, may show banding patterns which follow the rules discussed above, and listed in Figure 14, for constazyme patterns. If all of the subunits in a molecule are identical prior to the mutation, a dimer molecule, if the subunits combine randomly, may show 3 bands, a tetramer 5 bands, etc. The situation becomes complex, but interpretable, if a molecule already consisting of two types of subunits undergoes mutation affecting one subunit. For example, in the LDH system, a mutation producing a heterozygous

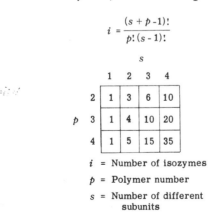

$$i = \frac{(s+p-1)!}{p!(s-1)!}$$

		s			
		1	2	3	4
	2	1	3	6	10
p	3	1	4	10	20
	4	1	5	15	35

i = Number of isozymes
p = Polymer number
s = Number of different subunits

Figure 14. Formula and table for calculating the number of isozymes if the number of types of subunits and the number of subunits in each molecule are known. Reproduced with permission of Shaw (1964).

150 / Chapter 6. Present and Future of Isozymology

condition of the B subunit (Figure 13, slot 4) results in five types of LDH-1, four types of LDH-2, three types of LDH-3, two types of LDH-4, and shows the original single band of LDH-5 (because it has no B subunits). Thus, by studying genetic variants affecting isozyme systems it is possible to come to some rather good preliminary conclusions about the subunit nature of the molecule under study.

Besides patterns of this type, genetic variation may reveal information on whether or not certain isozymes share subunits. If a genetic variant results in changes in migration of one isozyme and not another, it can be concluded that the particular subunit showing variation is not shared. On the other hand, if two isozymes show differences in migration in concert with genetic variation, it can be concluded that they probably share the involved subunit. An example of this type was cited in Chapter 1, Section B, involving erythrocyte esterases (Tashian, 1965).

2. The Use of Isozyme Techniques in Population Genetics

The great contribution of isozyme techniques in the study of genetic variation in natural populations was briefly discussed in Chapter 1, Section B. As pointed out, it is possible to obtain a relatively unbiased sampling of the genome (by selecting enzymes) with this approach. And, of course, the small amount of crude, nonpurified, material required, and the relative simplicity of the procedures, facilitates population screening. A single fruit fly, for example, can be evaluated with ease. The results of studies of natural populations, as pointed out earlier, have shown that 30% or more of loci are polymorphic (Shaw, 1964, 1965; Lewontin and Hubby, 1966; Harris, 1966).

An explanation of the forces initiating and maintaining the vast amount of genetic diversity revealed by the isozyme studies cited above is one of the major challenges in population genetics today. There are a multitude of subquestions. Are all of the polymorphic genes currently under selective pressures, or are some "neutral"? What proportion are under selective pressure only at restricted periods of ontogenetic development? What proportion are under external environmental selective pressure? What proportion are under epistatic selective pressures, that is, are selected by pressures operating internally?

It seems likely that isozyme techniques will play a major role in the further study of such questions. There is no other current method which yields information on the genetic structure of a population to rival that obtained from isozyme studies. In future experiments dealing with natural selection, the isozyme methods would seem to offer similar advantages. A number of potentially profitable experimental approaches came to mind. Environmental contrasts, for example, can be devised. Do water populations with possibly less external environmental variation show less genetic variation, as measured by isozyme variation, than land populations? Specifically, do deep sea populations show the least variation of all? Some concept of the amount of pressure exerted by environmental variation can be obtained from such studies. Of course, with those isozymic systems in which a selective system can be suspected because of the nature of the catalytic function of the enzyme, it may be possible to devise more direct experiments to define the role of environmental variation, and test the strength of selective pressures necessary in maintaining genetic variation. An evaluation of the amount of epistasis involved in maintaining genetic variation can be obtained, perhaps, by the type of studies currently underway by Hoyt, Sing, and Brewer (unpublished), in which covariation in frequencies of alleles of metabolically related isozymes is being studied.

G. Substrate Specificities and Biochemical Relationship of Enzymes

Isozyme techniques offer unparalleled opportunities to conveniently study the range of substrate specificities of specific enzymes and to detect cross specificities of apparently otherwise unrelated enzymes. Prior to this approach, it was necessary to carefully purify enzymes to test substrate specificities. By varying electrophoretic migration by means of varying electrophoretic conditions, such as pH, it is possible to tentatively conclude that two or more substrate specificities reside in the same isozyme molecule, and not in two molecules which by chance have migrated to the same position (unless the two molecules are bound in a complex). Of course, the use of genetic variation in such studies is better evidence, i.e., if the electrophoretic variant has both substrate specificities, then the molecule probably does have both activities. The results of such studies have shown that usually so-called "specific" isozymes, such as LDH and alcohol dehydrogenase, are capable of acting on several structurally related compounds. Some hexokinases of *Drosophila*, for example, are active with both glucose and fructose (Figure 12, parts A and B) (Knutsen et al., 1969).

Zymograms also often show interesting substrate relationships between groups of isozymes. In hexokinase zymograms of *Drosophila*, four genetically independent groups of isozymes are capable of acting on glucose (Knutsen et al., 1969). It will be of interest in such cases to see if the enzymes are structurally related and have all evolved from a single primordial gene, or if evolution has evolved two or more distinct, unrelated proteins to catalyze overlapping reactions. Of course, as so-called "nonspecific" isozymes are considered, such as the esterases and phosphatases, the activities of which are revealed by synthetic substrates, one may fairly ask, if the cross specificities of isozymes revealed on such zymograms have any biological meaning, i.e., does the activity revealed have anything to do with *in vivo* function? It is a difficult question to answer at this point, but it would not be surprising to the author if isozymes which have overlapping activities against even a synthetic substrate would also have overlapping activities *in vivo*, and possibly even be evolutionarily related.

H. Plant Isozymes

Much of the recent work with plant isozymes has been nicely reviewed by Scandalios (1969) and by Shannon (1968). The review by Scandalios covers amylase, catalase, alcohol dehydrogenase, leucine aminopeptidase, and peroxidase in some detail. The review by Shannon is short but lists 256 references to plant isozymes.

Work with plant isozymes, for some time carried out by only a few laboratories, has begun to become much more popular as larger numbers of biologists, botanists, systematists, plant breeders, and others have begun to realize the power of the technique. In general, the findings with plants have been similar to the findings with animals, i.e., genetic variation is prevalent in outbreeding populations, there is marked tissue–organ and ontogenetic variation in isozymes patterns, and there are many constazymes (nonsegregating) isozymes present in a high proportion of the systems. In many cases, the use of plants may offer considerable advantage over animals for the study of specific problems. If greenhouse space is available, many plants can be raised, reproduced, crossed, etc., in relatively little space, cheaply, and without disagreeable odors. Ontogenetic and organ studies are also usually relatively easy to carry out.

Beyond this, plants may provide unique material for study of certain problems. For example, we have used wheat species in our laboratory to study evolution of enzymes (see also Section I of this chapter). Wheat forms a polyploid series of diploid, tetraploid, and hexaploid species. The phylogenetic relationships of the development of the polyploid species has been reasonably well worked out. Contemporary species at all ploidy levels are available. Further, E. R. Sears of the University

I. The Use of Isozymes for the Study of Evolution / 153

of Missouri (Sears, 1966) has formed nullisomic-tetrasomic combinations of one of the hexaploid wheats so that the contribution of each chromosome can be evaluated. This kind of material does not exist in the animal kingdom. In the next section of this chapter (Section I) the use of this unique material in the study of evolution is examined.

I. The Use of Isozymes for the Study of Evolution

Many of the foregoing sections have dealt more or less with evolution, but in this section we wish to give some additional emphasis to the use of isozyme techniques in the study of evolution.

A number of examples could be cited. LDH has been used by a number of laboratories to study evolutionary relationships. Tashian's laboratory has effectively used carbonic anhydrase for the same purpose. In our laboratory an organism (wheat) has been selected and a multi-system approach used. Because of our experience with wheat, we will use it as an example of the use of isozymes in the study of evolution.

Wheat is a self-pollinating plant; this feature makes it of interest in constrasts with out-breeding organisms. Complete self-pollination does not permit an evolutionary strategy of genetic polymorphism. Presuming that gene product multiplicity is a selective advantage, an inbreeding organism is restricted to protein multiplicity of the constazyme type, common to all members of the population. Such multiplicity presumably arises in inbreeding organisms from gene duplcation and polyploid evolution followed by mutation. In this situation, we might anticipate that an organism such as wheat would have a considerable amount of isozyme multiplicity of the constazyme type. It was of interest, therefore, to observe that 100% of isozyme systems studied in wheat demonstrated more than one band of activity (Sing and Brewer, 1969), compared to about 60% for man; this suggests that wheat has indeed utilized an evolutionary strategy of gene duplication to obtain protein multiplicity.

The probable evolutionary history of wheat is shown in Figure 15. In general, the hexaploids have the widest geographic distribution, followed by the tetraploids and the AA diploids, with the BB and DD diploids having the narrowest distribution. Sing and Brewer (1969) have tested the hypothesis that the polyploids owe their greater ecogeographic distribution to a greater amount of protein multiplicity. The concept often put forth is that the duplicated loci in the extra genomes of the polyploid have a chance to "experiment" by undergoing mutation. With such a hypothesis, the number of each type of gene product in

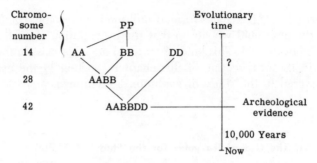

Figure 15. Probable evolutionary history of wheat. A primordial 14 chromosome wheat, indicated by "PP," evolved into various diploid species, indicated as "AA," "BB," and "DD." "AA" and "BB" gave rise to a tetraploid, and then, approximately 10,000 years ago, the tetraploid hybridized with the "DD" diploid to give rise to the "AABBDD" hexaploid, which is the commonly cultivated wheat grown around the world.

the polyploid should be increased, on the average. However the average number of isozymes in wheat seed material showed no significant differences between the AA, AABB, and AABBDD species (Sing and Brewer, 1969). All three of these species had a significantly greater mean number of bands than the BB and DD species. In leaf material there were no significant differences among any of the species. The specific isozyme patterns among the species were, in general, quite different from each other.

Brewer *et al.* (1969) then studied the contribution of each chromosome to the isozyme pattern in Chinese Spring hexaploid wheat by using the nullisomic–tetrasomic material developed by Sears (1966)

	Nulli-tetra series		
Genome →	AA	BB	DD
	0	1111	11
	0	11	1111
Homologous	1111	0	11
Group No. 1	11	0	1111
	1111	11	0
	11	1111	0

No. 2, etc.

Figure 16. The six possible nullisomic–tetrasomic combinations of hexaploid wheat involving chromosome group number 1. An equal number of combinations can be generated for each of the other six chromosome groups, making 42 possible combinations.

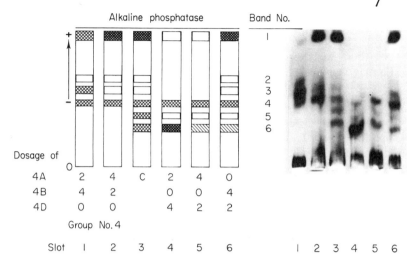

Figure 17. An electrophoretic study of wheat alkaline phosphatase isozymes involving nullisomic-tetrasomic chromosome group number 4 combinations. (Photograph on the right, explanatory diagram on the left.) Material from a control hexaploid is present in the third slot, marked "C." The dosage of each of the number 4 chromosome groups is indicated below each slot. It is apparent that the absence of the 4D genome (slots 1 and 2) causes disappearance of bands 5 and 6, while absence of the 4B genome (slots 4 and 5) causes disappearance of bands 1, 2, and 3.

(Figure 16 illustrates the various chromosomal combinations available for the first chromosome group). In the case of alkaline phosphatase, a contribution of each chromosome of the number 4 group could be detected, indicating that a locus for alkaline phosphatase was present on chromosome group number 4, and that each locus was different than the others (Figure 17). Surprisingly, eleven other isozyme systems showed no difference among any of the plants, indicating that within the limits of the electrophoretic method, the other loci are essentially identical. Since the contemporary diploids all show different isozyme patterns, it was surprising that the diploid genomes embedded in the hexaploid were similar at eleven out of twelve loci.

To demonstrate that recently synthesized polyploids would show parental isozyme patterns, Sing et al. (1969) have studied such material. As a general rule, each parental contribution can be detected in recently synthesized polyploids; this also means, of course, that the latter have more bands than the parents.

These studies with wheat may have far-reaching evolutionary implications. They indicate that, contrary to earlier concepts, the polyploid plants do not "experiment" with most of their duplicate genes, resulting in great gene product multiplicity, but, rather, they may tend to "converge" their variant genes. The mechanism of this remains to be investigated, but one possibility is chromosomal rearrangement through an occasional failure of homoeologous pairing. Certain genes may "resist" converging and remain (or become) divergent because of specific selective advantage. In this case, a variety of molecular forms of alkaline phosphatases may be an advantage in soils of varying phosphate content. Since polyploid evolution has probably played a key role in the general biological evolution of both plants and animals, its study is potentially quite important in understanding evolution.

Another approach to the study of the role of gene product multiplicity as represented by isozymes in evolution is to attempt to establish a relationship between the number of isozymes and the tissue complexity of an organism. Since one role of isozymes is, apparently, to relate to tissue differentiation, more complex organisms might be expected to have a greater amount of isozyme multiplicity. In a study of plants ranging in complexity from bacteria to corn, an increasing number of bands was seen, in general, as the organisms increased in complexity (Brewer and Sing, 1968).

Besides tissue differentiation as a driving force for isozyme multiplicity, intracellular diversity (as discussed in Section D, this chapter) may also be involved. Sing and Brewer (unpublished observations) have studied correlations of isozyme multiplicities in ten enzyme systems across twenty-two widely divergent plant species. It was of interest that those enzymes which were closely related metabolically tended to have high correlations of multiplicity. That is, there tends to be a correlation across species in the number of enzyme bands in metabolically related enzymes. It is possible that one driving force for multiplicity is the need for a variety of intracellular enzymes for intracellular organization purposes. For example, if one G-6-PD isozyme is localized in the cytoplasmic mitochondria, and another in the nucleus, efficiency of organization might call for the next enzyme in that particular pathway, 6-phosphogluconate dehydrogenase, to have one enzyme each for these intracellular locations. If this is the case, then it may be that the primary purpose of a portion of the structure of an enzyme protein is involved with positioning it intracellularly at the right organelle in association with the right sister enzymes. Of course, if a whole series of enzymes in a metabolic pathway have originated from gene duplication or polyploid evolution followed by mutation, they may have quite similar primary structures in that part of the enzyme molecule involved in locating the enzyme intracellularly.

REFERENCES

Brewer, G. J., and Sing, C. F. (1968). *J. Clin. Invest.* **47**, 11a.
Brewer, G. J., Sing, C. F., and Sears, E. R. (1969). *Proc. Natl. Acad. Sci. U. S.* (in press).
Cahn, R. D., Kaplan, N. O., Levine, L., and Zwilling, E. (1962). *Science* **136**, 962.
Davidson, R. G., Nitowsky, H. M., and Childs, B. (1963). *Proc. Natl. Acad. Sci. U. S.* **50**, 481.
Eaton, G. M., Brewer, G. J., and Tashian, R. E. (1966). *Nature* **212**, 944.
Emerson, P. M., and Wilkinson, J. H. (1965). *J. Clin. Pathol.* **18**, 803.
Gartler, S. M., Ziprowski, L., Krakowski, R. E., Szeinberg, A., and Adam, A. (1966). *Am. J. Human Genet.* **18**, 282.
Harris, H. (1966). *Proc. Roy. Soc. (London)* **B164**, 298.
Knutsen, C. A., Sing, C. F., and Brewer, G. J. (1969). *Biochem. Genet.* **3**, 475.
Latner, A. L., and Skillen, A. W. (1968). "Isoenzymes in Biology and Medicine." Academic Press, New York.
Lewontin, R. C., and Hubby, J. L. (1966). *Genetics* **54**, 595.
Linder, D., and Gartler, S. M. (1965). *Science* **150**, 67.
Markert, C. L. (1962). *In* "Hereditary, Developmental, and Immunological Aspects of Kidney Diseases" (J. Metcaff, ed.), p. 54. Northwestern Univ. Press, Evanston, Illinois.
Markert, C. L. (1963). *Science* **140**, 1329.
Migeon, B. R., and Miller, C. S. (1968). *Science* **162**, 1005.
Prout, G. R., Macolalag, E. V., and Hume, D. M. (1964). *Surgery* **56**, 283.
Richterich, R., Locher, J., Zuppinger, K., and Rossi, E. (1962). *Schweiz. Med. Wochschr.* **92**, 919.
Scandalios, J. G. (1969). *Biochem. Genet.* **3**, 37.
Sears, E. R. (1966). *In* "Chromosome Manipulations and Plant Genetics" (R. Riley and K. R. Lewis, eds.), pp. 29–45. Oliver & Boyd, Edinburgh and London.
Selagi, S., Darlington, G., and Bruce, S. A. (1969). *Proc. Natl. Acad. Sci. U. S.* **50**, 481.
Shannon, L. M. (1968). *Ann. Rev. Plant Physiol.* **19**, 187.
Shaw, C. R. (1964). *Brookhaven Symp. Biol.* **17**, 117.
Shaw, C. R. (1965). *Science* **149**, 936.
Shaw, C. R., and Barto, E. (1963). *Proc. Natl. Acad. Sci. U. S.* **50**, 211.
Sing, C. F., and Brewer, G. J. (1969). *Genetics* **61**, 391.
Sing, C. F., Brewer, G. J., and Sears, E. R. (1969). In preparation.
Starkweather, W. H., Green, R. A., Spencer, H. H., and Schoch, H. K. (1966a). *J. Lab. Clin. Med.* **68**, 314.
Starkweather, W. H., Spencer, H. H., and Schoch, H. K. (1966b). *Blood* **28**, 860.
Tashian, R. E. (1965). *Am. J. Human Genet.* **17**, 257.
Weiss, M. C., and Green, H. (1967). *Proc. Natl. Acad. Sci. U. S.* **58**, 1104.
Wilkinson, J. H. (1965). "Isoenzymes." Lippincott, Philadelphia, Pennsylvania.
Wroblewski, F., and Gregory, K. (1961). *Ann. N. Y. Acad. Sci.* **94**, 912.

Chapter 7

Analysis of Electrophoretic Variation[a]

A. Introduction

In this chapter we will consider some aspects of acquiring data from electrophoretic runs and using the data to extract meaningful statements about the phenomenon being studied. For the most part, this chapter is intended for the investigator, although the clinician or clinical pathologist may find certain sections useful, particularly when large scale screening applications are involved.

The foremost consideration in all studies is the research plan. This begins with a clear statement of the goals of the research to be done. Next one sets down the methods to be used to acquire and analyze the data. In this respect, an early decision in any study is whether one has the resources to collect and analyze an adequate sample to answer the question at hand. The scope of the study may be limited by the time and money at the disposal of the researcher. In some instances it may be wise to delay until adequate time and funds are available. It is not that all good enzyme studies must be extensive. It is simply that if there is considerable biological and environmental variation, a large sample from which to generalize is required.

The second general question one should be able to answer is whether the sample collected by the study will be representative of the biological material about which the investigator wishes to make an assertion. This often involves good judgment and adequate experimental design (see Section C, 4, a). Given that the sample is representative, we may further ask how reliable our statements surmised from the sample will be in

[a] Contributed in major part by Charles F. Sing, Department of Human Genetics, University of Michigan.

predicting the true (unknown) characteristics of interest. Statistics enter the picture here (see, in general, Section C, this chapter).

B. Data Acquisition

1. Introduction

The objective of scoring the results of an electrophoretic run is to obtain a permanent record for reference or analysis. Photographs are one kind of record. They are often desirable for reference or publication but fall short, first, in reflecting all the information which can be extracted (such as quantitation) and second, in presenting the data in a form suitable for statistical analysis.

Obviously the type of information a researcher takes from a gel is related to the question he is asking. The data could be as simple as noting the classification of the pattern types. This simple record of the pattern is likely to be adequate to estimate gene frequencies or detect individuals with a variant pattern in a survey. On the other hand, to satisfy the requirements for certain studies, the record for an individual could be as complex as the graphic display of the optical density across the gel pattern. With the data record in this form one could retrieve the number of bands (peaks), their relative positions, and the fractional contribution of each band to the total quantity of enzyme (area under segments of the curve).

In any case, one should define as much as possible, before the experiment begins, what he will measure and how it will be done. In doing so, questions will emerge such as what constitutes a band, what constitutes an "A" pattern, or on what scale will the density of a band be recorded. The result will be a more consistent score which will depend less on the individuality of the scorer. Scoring of the electrophoretic result should be as consistent as possible from run to run. Likewise, the conditions under which measurements will be taken can play a large role in obtaining meaningful data and should be standardized wherever possible. Inconsistency in the scale on which the patterns are scored (including standardization of measurement devices) or any condition which affects measurement repeatability may tend to increase experimental error. Obvious extraneous variables such as voltage, temperature during electrophoresis and starch lot should be held as uniform as possible to control variation which might result among patterns when these factors are ignored. A careful definition of the variable being measured and consideration of extraneous environmental effects associated with the study will pay big dividends when one sits down later to draw inferences from the analysis of the data. The larger the error variation, the greater the difficulty in making inferences about the phenomenon being studied. It is a stronger approach to increase the precision of measurement to reduce errors than to use statistical methodology to remove extraneous sources of variability from the data after it is collected.

In making these planning decisions one can increase the usefulness of the data as a future reference if the record is as comprehensive as is feasible. Experience suggests that an investigator will return to his data to ask questions other than those originally intended. For this reason, an auxillary consideration in acquiring and recording data should be to obtain the maximum amount of information allowable under the constraints of time and money allocated for the study.

Once the type of data which is to be collected is ascertained, the manner in which the information will be stored, the format of the record, becomes important. In this respect the form the record takes should be ascertained in terms of its ease in manipulation to answer immediate questions and its accessibility for future reference. Sorting out information from handwritten records is, in most situations, more time consuming than sorting from coded machine records such as punched cards. It is apparent that an inefficient record system would tend to be more limiting as the volume of data increases. With large scale surveys or experimental situations with many replicated treatments, storing data in an easily retrievable form becomes essential.

We will first discuss the modes of taking the data and placing it in a record form (Section B, 2). We will then discuss the manipulation of the data record to enhance accessibility (Section B, 3).

2. Methods of Recording Electrophoretic Data

a. Sample Small, Simple Notebook Records

When one has a small number of gel patterns, a descriptive statement recorded in a notebook is quite often satisfactory. Examples would be preliminary studies to establish the techniques for a new enzyme system and the effects of certain manipulations on the enzyme pattern (such as changing temperature or length of the run).

For established enzyme systems, the record could simply be the designation for the genotype or phenotype known to be associated with the pattern. Intermediate to simple written records and photographs is the drawing of a sketch in the notebook to represent the pattern.

b. Sample Small, Photographs

Photographs can act as an excellent ancillary to the description of the outcome of a gel run. Especially useful is the Polaroid system. Photographs can be taken by this system with black and white or color film and developed immediately, enabling one to observe the outcome when the run is still fresh. Of course, this allows the advantage of being able to revise the lighting or staging of the gel so that the picture portrays the material as closely as possible.

c. Sample Large, Simple Note Taking

Electrophoretic data is increasingly being collected on large samples of individuals or experiments involving the comparisons of many types of biological material. In these cases a body of data can be generated which may exceed the feasibility of photographs. There are three types of information which can be obtained from a gel pattern: (1) the number of bands, (2) the density of each of the bands, and (3) the relative positions of the bands. Each of the variables might be examined visually and recorded. An example of a record sheet used in a large study that was designed to include information on each band is given in Figure 18. This sheet was constructed to serve as a permanent or intermediate record in obtaining a punched card record. The variable of interest is recorded in a position on the sheet which corresponds to each band. It is relatively simple to record the data according to a code which corresponds to characteristics of the band (light, heavy, broad, narrow, etc.). A measurement on a scale which reflects the relative position of the band may also be entered for each band from, say, the anodal to the cathodal ends of the gel.

d. Sample Large, Automated Machine Records

As the amount of information taken from each pattern increases, so does the burden of retrieving information from the written record. At a certain point the number of pieces of information which can be tabulated and summarized efficiently by inspection exceeds feasibility and efficiency. One usually then turns to automated machine records. Typically data collection of this type requires three steps. The data is first recorded on some type of score sheet, transferred to punch cards, then fed into a computer. However, a number of variations are possible. This topic will be discussed in detail in Section C.

e. Densitometry

Densitometric methods for analysis of the amount of proteins on electrophoretic media are fairly well established. In general the amount

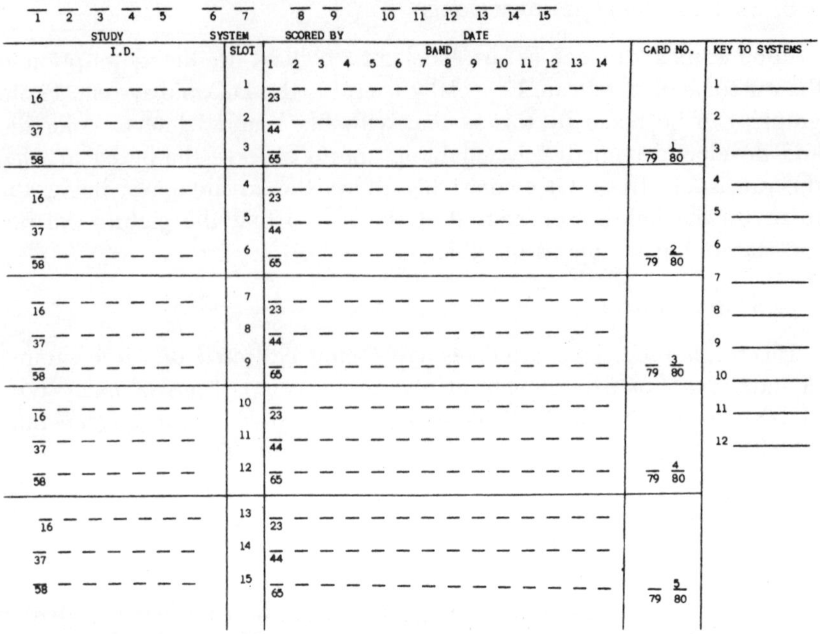

162 / Chapter 7. Analysis of Electrophoretic Variation

Figure 18. An example of the sheet used to record data from 15 slots of an electrophoretic run. The number below each character of information is the column number in which the data is entered on punched cards.

of protein in a band correlates reasonably well with the amount of protein stain present, although some proteins have a higher affinity for the stain than others, which may, under certain conditions, make them stain relatively more intensely than they should.

The problems are greatly multiplied in attempts to quantitate enzyme bands on electrophoretic media after histochemical staining. In contrast to protein stains in which the end result is essentially a stoichiometric conversion of dye per given amount of protein, the important factor with enzymic quantitation is activity, which of course is measured by rate of reaction. Rate must be determined during the reaction and not after its completion; thus timing is of great importance. The reaction must be stopped before substrates and cofactors are depleted, and before intensity reaches a maximum at any one site. To make rates from one run to another, comparable conditions must be kept very standard; this is difficult to do in electrophoretic media. Variation in substrate and cofactor affinities of various isozymes is a further problem. Optimal substrate and cofactor concentrations for one isozyme may not be appropriate for another.

B. Data Acquisition

Some isozymes tend to have broader bands than others, which will make the staining intensity vary. Some stains seem to have a threshold, i.e., a certain threshold number of molecules of dye must be converted before the dye turns color and/or precipitates. If the reaction is very fast in some bands the end product of the reaction may be developed more rapidly than dye molecules became available, allowing part of the end product to diffuse away. If a precipitate of the dye forms on the gel surface by virtue of enzyme action, it may limit diffusion of more substrate further into the gel. Varying thickness of the gel is another potential problem.

The above problems with quantitation of histochemical stains on electrophoretic media suggest that these methods should not be undertaken lightly. On the other hand, it is our view that improvements in equipment, and the need for such objective measures, will increasingly

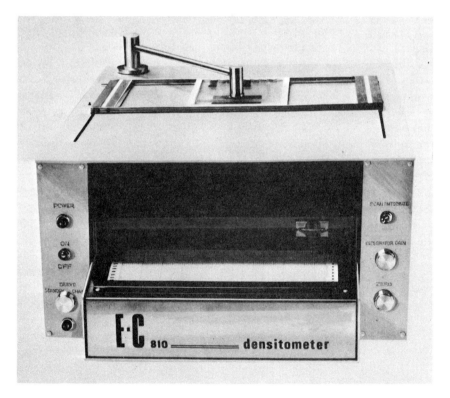

Figure 19. The E.-C. 810 optical densitometer manufactured by the E.-C. Apparatus Corp., Philadelphia, Pennsylvania. Photograph provided by courtesy of E.-C. Apparatus Corp.

make these methods more desirable. The remainder of this section discusses the most applicable of presently available equipment.

Two basic methods may be used: reflectance and densitometric. If the media is completely impermeable to light, a reflectance method must be used. This would apply particularly to paper electrophoresis. With less opaque media, densitometric methods involving a measurement of the amount of light transmitted through the media can be used. The latter method is considerably simpler and probably more accurate. Acrylamide and agar gels and cleared cellulose acetate strips can be used quite well. With reasonably high intensity light sources, such as with the Gilford equipment described later in this section and also in Section B, 3, starch gels can also be used, even without clearing. We have had no experience with the use of the E.-C. densitometer on starch gels.

An optical densitometer, such as the E.-C. 810 manufactured by the E.-C. Apparatus Corp., Philadelphia, Pennsylvania (Figure 19), can provide a linear reading over the optical density range from 0–2.0 absorbency. The linear response with density, according to Beer's law, allows one to interpret the integral as a measure of the amount of stain in the peaks of the integram. Essentially it is providing a quantitative (and qualitative indication of the peaks) assay of the presence of materials on the gel which vary in their optical absorption characteristics from that of the gel matrix. The measurement comprises a comparison of transmitted light intensity compared to some reference area, usually some unstained section of the gel. A schematic drawing of the operation of the model 810 is given in Figure 20. Although instruments may vary in terms of their designed ability (filtering, wave length variations, and sensitivity to external perturbations), all optical densitometers are based on this one principle. A general discussion of optical methods based on the analysis of photochemical absorption is given by Kay (1964).

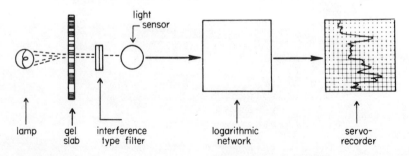

Figure 20. A schematic diagram of the operation of the E.-C. 810 optical densitometer.

The E.-C. 810 is designed to assay an intact standard acrylamide gel slab. It can be converted to accept other media. The Gilford 2400 system, manufactured by Gilford Instrument Co., Oberlin, Ohio, has a sampling compartment which will handle a variety of cuvettes. One such cuvette will accommodate gel cylinders and slabs up to 8.5 mm wide $\times$ 100 mm long. Most gel slabs would have to be sectioned for analysis on this machine as it is presently designed.

3. Storing and Handling the Record by Machine Methods

Regardless of whether the record is the result of a visual evaluation or the output of an optical instrument, one is usually faced with having to retrieve the record one or more times before the study is completely analyzed. Simply sorting through the written records for certain pieces of information is a very time consuming and tedious operation. Methods have been developed to use a record which can be sorted and tabulated by machine. The machines (hardware) available for such jobs range from inexpensive card sorters to the giant computers which have become the centers for monitoring large complex industrial operations. In addition to the increased speed at which the records can be manipulated, there is an increase in accuracy of tabulation (the search and recording of certain characteristics of the data is not affected by human error).

The usual procedure is to enter the data onto a record sheet (possibly such as illustrated in Figure 21) in some sort of short or coded form. The object of this first record is to provide a means to assemble the information collected from a study in a form which can be transferred *in toto* at a convenient time to machine records. Several methods of "accessing" the data from this first record to the machine are available.

The most widely used automated record is the punched card. The data are punched onto the cards from the record sheet. An example is given in Figure 21. The punched card is one of the most flexible and economical machine records. It may be sorted and tabulated with a card sorter such as the IBM 082. For more complex computations it may be used as the input to a high speed digital computer. If one has large enough volume to justify the rental or purchase of a key punch for the laboratory, it may be desirable in certain situations to bypass the first written record sheet and punch the information directly from the gel run.

Written records may also be bypassed by incorporating devices which automatically punch cards or a paper tape, as the gel is scanned. An illustration of deriving punched tape records from spectrophotometer is given in Figure 22. This is configuration of instruments available

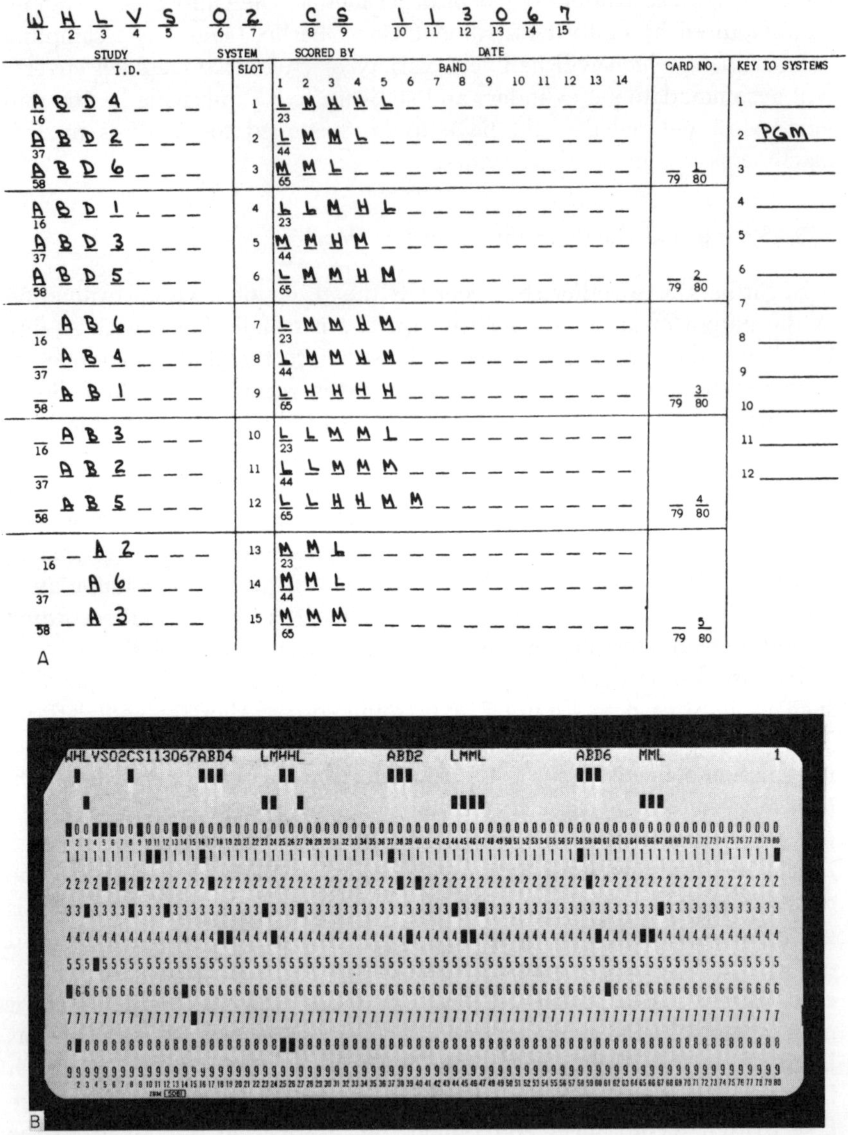

166 / Chapter 7. Analysis of Electrophoretic Variation

W H L V S	O 2	C S	1 1 3 0 6 7		
1 2 3 4 5	6 7	8 9	10 11 12 13 14 15		
STUDY	SYSTEM	SCORED BY	DATE		
I.D.	SLOT	BAND		CARD NO.	KEY TO SYSTEMS
		1 2 3 4 5 6 7 8 9 10 11 12 13 14			
A B D 4 / 16	1	L M H H L / 23			1 _____
A B D 2 / 37	2	L M M L / 44			2 PGM
A B D 6 / 58	3	M M L / 65		79 1 / 80	3 _____
A B D 1 / 16	4	L L M H L / 23			4 _____
A B D 3 / 37	5	M M H M / 44			5 _____
A B D 5 / 58	6	L M M H M / 65		79 2 / 80	6 _____
					7 _____
_ A B 6 / 16	7	L M M H M / 23			8 _____
_ A B 4 / 37	8	L M M H M / 44			9 _____
_ A B 1 / 58	9	L H H H H / 65		79 3 / 80	10 _____
_ A B 3 / 16	10	L L M M L / 23			11 _____
_ A B 2 / 37	11	L L M M M / 44			12 _____
_ A B 5 / 58	12	L L H H M M / 65		79 4 / 80	
_ _ A 2 / 16	13	M M L / 23			
_ _ A 6 / 37	14	M M L / 44			
_ _ A 3 / 58	15	M M M / 65		79 5 / 80	

A

Figure 21. Data recorded from a phosphoglucomutase gel on November 30, 1967 for the wheat leaves study (A). The punched card (B) represents the first three lines of information. Each band was scored L = light, M = medium, and H = heavily stained.

B. Data Acquisition / 167

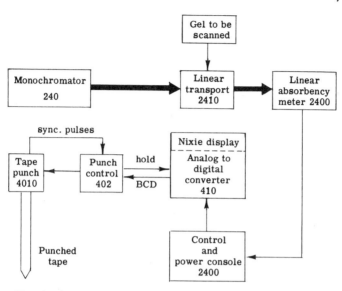

Figure 22. A schematic diagram of an instrument configuration designed around the model 2400 spectrophotometer manufactured by the Gilford Instrument Laboratories, Inc., Oberlin, Ohio.

from the Gilford Instrument Laboratories, Oberlin, Ohio. The object here is to convert the plot of linear absorbency across a gel pattern to a binary record punched in a paper tape (a card punch may be substituted here if card records seem more desirable). This binary tape record then serves as an input to a paper tape reader linked with a computing system. As indicated by the flow of information in Figure 22, an analog signal is first converted to decimal code in the 410 converter. The nixie display (so named because a nixie tube is employed) is a digital readout of the conversion as the information is processed. The decimal information is converted to binary form (BCD—Binary Coded Decimal) so that more information can be stored per unit of tape.

Large data collections may justify storing records directly onto magnetic tape or a magnetic disk. Several types of input devices are now available which enable the researcher to feed data directly from the laboratory to storage files incorporated in some central computing facility. Here the researcher would require "on line" access to the computing machinery with these files in a "ready state" whenever he was formulating his records. Except for a very large operation it is usually more economical to formulate an intermediate storage file which does not require computer time. The punched card and punched paper tape records form such storage files.

C. Data Reduction

1. Introduction

Many biologists while acquiring basic knowledge in various disciplines do not gain a strong background in biomathematics. This often leaves them unprepared to effectively reduce a large mass of observations to a meaningful form. Further, a superficial approach to data may allow many meaningful relationships to go undiscovered and too often, poor planning in data collection seriously handicaps the amount of information that can be gained from a study.

By data reduction we imply the summarization of the information in a sample in terms of a few numbers. In short, the objective of data reduction is to obtain reliable estimates of parameters which describe the population from which the data has been obtained. Since one objective of this book is to assist at all levels of investigations in collection and analysis of electrophoretic data, we will start with some basic statistical principles.

2. Statistics and Parameters

The data one collects is a "sample." The larger body of data from which the sample is drawn is called the "population." The population may be enumerable, but is quite often purely hypothetical in that it could include an infinite number of observations. We compute "statistics" from the sample to estimate "parameters" which describe the population. If the data collection is accurate and free from bias the statistics will be "good" estimates of the true unknown parameters.

The parameters which are often chosen as giving the most information about a given variable are the mean and the variance. The mean is the weighted average of the possible values the variable may take (each weighted by their respective frequency of occurrence). The spread of values about the mean is reflected in the variance. The variance also enables one to make a statement about the reliability of the mean as a prediction of any one observation drawn from the population.

In practice we do not know the true values of these parameters. The number of possible measurements is usually so large that it is not feasible to attempt to sample the entire population. Hence, we draw a sample of measurements to estimate the parameters of the population. This is termed point estimation. Other parameters which might be of interest from the standpoint of point estimation are the regression coefficient and the correlation coefficient. Both are measures of association between pairs of measurements taken on each sampling unit which estimate the true relationship between two variables.

Each of the "point" estimates has a corresponding variance which is also estimated from the data. The variance of each point estimate is a function of the variance of the sampled observations. In addition to providing a measure of how reliable our point estimate is, the estimated variance provides a methodology to judge whether one point estimate is inherently different from another or whether the observed difference is within the range expected by experimental error. This latter aspect of statistics is referred to as hypothesis testing.

Biostatistics (biometry) then, is a discipline concerned with (1) the reduction of data to meaningful estimates (statistics) of parameters which are of interest to biologists and (2) the tests of hypotheses which have been made concerning the variable of interest. Many fine texts (Steel and Torrie, 1960; Snedecor and Cochran 1967) are available which provide a comprehensive coverage of these two aspects of data reduction.

3. Origin and Meaning of Variation

The reason why a sample of observations does not always give a straightforward answer to a given question is variation. Variation in biological material and/or errors in characterizing the biological specimen, causes us to obtain different measurements from repeated observations of the population being studied. In the more exact sciences such as physics and chemistry, the irreducible variation may be very small, often no more than experimental errors. With biological variables, additional variation is inherent and must be accepted as a part of the nature of the material. Ignoring this variation can only contribute to making an erroneous inference from the experimental measurements. For example, we often use the mean to portray the most likely measurement even though it conceals much about the true nature of specific observations. By knowing the variance, we have some concept of how predictable the mean actually can be. Intuitively, we recognize that the larger the variance the less accurate the mean is as a predictor of any specific measurement that might occur on one observation. Conversely, in the limiting case of no variation, the mean would provide a perfect prediction of the outcome of any specific observation.

To illustrate the point, let us consider the distribution of the number of hexokinase isozymes in a population, of say, *Drosophila robusta*. Would the observations from one fly (a sample) of this species provide an adequate characterization? Certain statements could be made. A statement could be made about whether or not hexokinase isozymes are present in this species. Furthermore, one could define very carefully the hexokinase information on the gel pattern represented by the single fly. We could say much about the hexokinase of the fly under study, but how much can we say about the hexokinase of *D. robusta* in general? Granted, if we only had one fly this would be our best characterization of the hexokinase pattern of the species. The adequacy of this estimate to provide inferences for the entire species would depend on the variation in the pattern present in the population. If variation in hexokinase activity does exist, one fly cannot be a representation for the species in general. Variations may be due to genetically determined differences or to errors of measurement due to extraneous variables such as stage of development or pretreatment of the fly. To properly describe the hexokinases of the species we need to know the range of variation in patterns as well as the most likely pattern. This is done by collecting a sample of flies. In general, the greater the variation in the population, the larger the sample should be in order to characterize the population with reasonable confidence.

4. Possible Methods of Obtaining More Information from a Study

The application of the electrophoretic technique to problems such as the *Drosophila* problem mentioned above are becoming more common. Studies are being conducted to obtain a description of biochemical variation in a population of individuals and to determine the effects of experimentation on these enzyme variations. We have discussed the importance of a research plan and a method of data collection. We also pointed out (Section A) the importance of obtaining a sample which can be used to make meaningful statements about the population of interest. Here we will discuss ways the researcher can improve his study to achieve this latter goal.

We judge how meaningful a statement is about a parameter in terms of the accuracy of and information about the estimating statistic. Accuracy is measured in terms of how close the statistic approaches the true value. The information we have about a parameter is inversely related to the variance of the corresponding statistic. (Some authors prefer to consider a decrease in variance as an increase in "precision.") Both considerations fall into the general category of experimental design.

C. Data Reduction / 171

By deciding on an experimental design, one determines how the sample will be obtained from the population.

a. EXPERIMENTAL DESIGN

Experimental design is an attempt to manipulate the collection of data so as to obtain more reliable (small variance) and unbiased (accurate) statistics. The sensitivity of tests of hypotheses will also be increased by designing a sampling layout which will accomplish these objectives.

A word should be said at this point about the effects of bias, and sampling variance on hypothesis testing. Here we are concerned with sampling bias. It is the deviation of the observed statistic from the estimate which would obtain if the sample were collected in a way so as to be free of all systematic effects which could affect the statistic. The effect of bias on a statistical test is to make invalid the difference which is being tested. Otherwise stated, is the difference between the statistic and some other value due to sampling bias or a true difference? A discussion in Section C, 4, b will suggest methods to minimize sampling bias. The effect of reducing error variability in a study will be to increase our ability to determine whether the deviation of the statistic from some other value is real or is of the order of expected chance deviation.

In brief, the theory of hypothesis testing is as follows: The estimate of experimental error is used to give a probability of finding a difference as large or larger than that observed if the true difference is zero. If the probability of this difference occurring when there is no difference (the null hypothesis) is small (usually set at 0.05; called the size of the test) then we reject the null hypothesis. In doing so we accept the risk (a probability equivalent to the size of the test) of making an incorrect decision. That is, we may have rejected the null hypothesis when in fact it was true. This has been described as a Type I error, or error of the first kind. The probability of detecting a difference when in fact one does exist is called the power of the test. As we indicated above, the smaller the error variance the more critical the experiment will be in detecting a true experimental difference. The failure to reject the null hypothesis when it is false has been labeled a Type II error. In terms of hypothesis testing, our objective then is to design an experiment which results in a test of the null hypothesis which has a maximum power for the chosen size of the test.

In keeping with the format of this discussion we will not consider the mechanics of statistical testing. It has not been our intent here to give a comprehensive coverage of statistical methodology. This is done very adequately and with great elegance in many standard reference works which are available, such as Steel and Torrie (1960), Snedecor and Cochran (1967), Peng (1967), or Cochran and Cox (1962). However, we hope that some of the concepts discussed will allow the individual researcher, particularly the inexperienced one, to consider the advantages of good experimental design.

b. Replication, Randomization, and Local Control

There are at least three facets of experimental design. They are replication, randomization, and "local control." Replication involves the number of individuals of each type sampled. Randomization is a system of drawing the observations for the sample to avoid bias. Local control is the blocking, grouping, stratifying, or otherwise choosing more homogenous groups from which to draw the observations. All three strategies tend to increase precision. No estimate of experimental error is possible without replication. Replication may be repeated observations (several gel slots, for example) on the same biological specimen, or repeated observations of a population of organisms. In the former we are concerned with the error variance associated with the statistics computed from the observations on a series of samples from an individual; in the latter we are concerned with the error variance of statistics computed on a sample from a population. In either case, as the size of the sample increases, the error variance of the corresponding statistic decreases.

We randomize the collection of observations of a variable to eliminate the systematic effects extraneous variables could have on our study. The general principle is to eliminate any bias which might occur by balancing the effects of extraneous variation. We balance the extraneous variation by randomizing these effects among the observations which comprise the sample. As an example, consider the effect of ordering specimens in the slots of a starch gel slab. We would decrease gel edge bias or the opportunity of systematic scoring bias by randomizing the treatments on the gel in each run rather than placing them in some other order. If randomized, each specimen would have an equal chance of appearing in any one slot. Knowing that, say, individuals of Type 1 were always placed in the first slot could result in it being scored differently than if this knowledge was not available. Cochran and Cox (1962) describe the randomization as "insurance against disturbances that may or may not occur and that may or may not be serious if they do occur."

By local control we imply the design proper. We are concerned about the design of an experiment because we recognize the effects that extraneous variables may have on the observations of interest. We attempt to arrange the collection of data so that the effects of these concomitant variables can be accounted for and removed from the estimate of error variation. Essentially this can involve two strategies. First, we can standardize all extraneous variables throughout the collection of the experimental observations. Holding temperature constant for all runs is an example. Second, we can group subsets of the observations into groups in which the extraneous variable takes on a homogeneous value and then use the differences among blocks to estimate the effect of the extraneous variable. Each group could be a scorer. We reduce the error mathematically by the amount which would otherwise be due to the scoring variation if grouping were ignored.

Design is essentially a part, or refinement, of experimental technique. In our problem of evaluating electrophroretic variation we may be concerned with the effects of variations in temperature, length of run, the individual doing the scoring, or any other environmental nonexperimental concommitant variable which could be affecting the outcome of the electrophoretic run. A brief example is offered to clarify the operational aspects of design. Say we collect two large samples of individuals on which an an electrophoretic analysis is to be done. Suppose that the question of interest is: Do these two samples represent the same population of individuals? Let the variable of interest be the number of molecular forms of a particular enzyme in each individual. Assume the size of the sample requires making several runs and that we are unable to control various aspects of the run precisely. The knowledge that variation in runs would tend to affect the number of bands measured on individuals for each run suggests some sort of methodology, a design, which would minimize this effect in the study. This may be accomplished by representing both samples on each run. The comparisons between samples are then based on data pooled from within each run. The "confounding" effect of running time is removed by "blocking" the observations. Our estimate of the error variation is then made from observations among individuals within each run. This estimate will be free of the component of variation due to runs which would otherwise be present among individuals if blocking were ignored. For studies of this sort the electrophoretic method which enables one to run many individuals on one run is obviously superior. Slab techniques allow such a design. Disc electrophoresis could not provide such "unbiased" estimates of error variation because only one individual is run per gel.

5. A Specific Example

To illustrate an example of data reduction we will draw on a study of isozymes of a polyploid series of wheat reported by Sing and Brewer

(1969). The research plan began with the question of the existence of a correlation between ploidy and multiplicity of molecular forms of enzymes? Stated as a null hypothesis: Differences do not exist in multiplicity of an enzyme among species differing in ploidy. Since much is known about the nature of the polyploid series of wheat, wheat was chosen as the organism to test the hypothesis. To obtain a test of the hypothesis, point estimates of mean multiplicity and error variation for species groups differing in ploidy were needed.

The next consideration was to decide how to acquire a representative sample of the population of possible observations about which inferences were to be made. This population consists of the set of possible measurements of multiplicity which could be made if all enzymes were measured on each of the species groups representing the three ploidy levels (diploid, tetraploid, and hexaploid). A sample was selected in the following manner. Eight enzyme systems which had electrophoretic methods established for other organisms were chosen. This sample of enzymes was considered to be representative with regard to multiplicity because it was chosen at random with respect to the multiplicity an enzyme may have. Note that choosing more than one enzyme is essentially replication which will improve the precision of our estimate of average enzyme multiplicity for a species group. Next, six individual plants were chosen at random to represent each species group. This replication within species group provides a more reliable estimate of the mean multiplicity of an enzyme for a group than would the study of an individual plant, and gives an estimate of error variation which is attributable to intragroup differences. Note that replication and randomization enter into the sampling design at every possible point as ways to assure that a more representative characterization of the population is being made.

Experimental design enters again in running the electrophoretic assays. First, "local control" is employed by developing a 30 slot gel which allows comparisons of all individuals on the same run. Any run to run (gel to gel) differences which could affect the banding patterns was removed by blocking all individuals into a homogeneous experimental unit. Second, the 30 individuals were assigned to the 30 slots of the gel in a random fashion. This was done to guard against scoring bias which might enter in if the scorer was aware of the positions of the species groups. Randomization also tends to randomize the individuals with respect to extraneous conditions associated with different positions across the gel, such as edge effect.

The data collected is summarized in Table IV. Each species group mean is based on 48 observations (6 individuals for each of the 8 enzyme systems). Likewise, each enzyme mean is the result of 30 observations. An analysis of the data to determine if there are significant differences among the species group means was done to test the null hypothesis. The hypothesis testing procedure used is generally described as the analysis of variance (see Peng, 1967, for details). The null hypothesis was not rejected when the size of the test was taken to be 0.01. The authors conclude that no significant differences in enzyme multiplicity were detected in this sample.

Table IV

SUMMARY OF THE AVERAGE NUMBER OF ISOZYME BANDS PER SLOT FOR WHEAT LEAVES

Species group	Phosphoglucomutase	Hexokinase	Acid phosphatase	Esterase	Malate dehydrogenase	Alkaline phosphatase	Glucose-6-phosphate dehydrogenase	6-Phosphogluconate dehydrogenase	Mean
AA	3.41	3.00	3.75	8.83	5.83	3.83	3.00	2.00	4.21
BB	3.75	1.50	2.58	10.08	5.50	5.40	3.25	2.41	4.31
DD	3.50	1.75	2.58	9.00	5.00	6.91	3.58	3.00	4.41
AABB	5.08	1.50	2.83	4.83	5.91	4.16	3.00	2.50	3.73
AABBDD	4.25	1.00	2.25	4.83	6.66	2.18	3.00	2.00	3.27
Mean	4.00	1.75	2.80	7.51	5.78	4.50	3.17	2.38	3.99

REFERENCES

Cochran, W. G., and Cox, G. M. (1962). "Experimental Designs." Wiley, New York.
Kay, R. H. (1964). "Experimental Biology." Reinhold, New York.
Peng, K. C. (1967). "The Design and Analysis of Scientific Experiments." Addison-Wesley, Reading, Massachusetts.
Sing, C. F., and Brewer, G. J. (1969). Genetics 61, 391–398.
Snedecor, G. W., and Cochran, W. G. (1967). "Statistical Methods." Iowa State Univ. Press, Ames, Iowa.
Steel, R. G. D., and Torrie, J. H. (1960). "Principles and Procedures of Statistics." McGraw-Hill, New York.

Author Index

Numbers in italics refer to the pages on which the complete references are listed.

A

Adam, A., 141, *157*
Aebi, H., 113, *134*
Ajmar, F., 91, *135*
Allen, J. M., 93, 94, *134*
Allen, S., 5, *14*
Anstall, H. B., 105, 106, *134*
Aurell, B., 51, *52*
Aw, S. E., 45, *52*, 133, *134*

B

Baker, R. W., 3, *14*, 41, *52*
Baron, D. N., 97, 98, *135*
Barron, K. D., 122, *135*
Barto, E., 10, *14*, 86, *137*, 147, *157*
Baughan, M. A., 75, 77, *134*, *135*
Bearn, A. G., 4, *15*, 116, *137*
Beck, C. C., 119, 121, *135*
Beckman, L., 100, 101, 102, *134*
Bell, J. L., 97, 98, *135*
Bergstrom, E., 18, *39*
Bernsohn, J., 122, *135*
Bertland, L. H., 117, *136*
Beutler, E., 73, 79, *136*
Bier, M., 6, *14*
Bigley, R. H., 106, *136*
Bjorling, G., 101, 102, *134*
Blanchaer, M. C., 44, *52*
Bloemendal, A., 51, *52*
Bloemendal, H., 33, *39*
Bodansky, O., 132, *136*
Bodman, J., 43, *52*
Bodmer, W., 6, *14*
Boivin, P., 18, *39*
Bourne, G. H., 94, *136*
Bourne, J. G., 88, *135*

Bowbeer, D. R., 82, 91, *135*
Bowman, J. E., 91, *135*
Boyer, S. H., 5, *14*, 92, 94, *135*
Brewer, G. J., 70, 73, 75, 77, 79, 82, 83, 84, 91, 94, 95, 97, 113, 115, 119, 121, 133, *134*, *135*, *136*, *137*, 141, 145, 146, 151, 153, 154, 155, 156, *157*, 174, *175*
Bruce, S. A., 144, *157*
Bucher, T., 116, *135*

C

Cahn, R. D., 86, *135*, 147, *157*
Campbell, D. M., 97, 98, *135*
Carson, N., 112, *136*
Carson, P. E., 91, *135*
Childs, B., 6, *14*, 73, *135*, 143, *157*
Christodoulou, C., 101, 102, *134*
Ciotti, M. M., 116, *136*
Cochran, W. G., 169, 172, *175*
Collier, H. O. J., 88, *135*
Cortner, J. A., 116, 117, *135*
Cox, G. M., 172, *175*

D

Daams, J., 18, *39*
Darlington, G., 144, *157*
Davidson, R. G., 6, *14*, 73, 116, 117, *135*, 143, *157*
Davis, B. D., 98, *135*
Davis, B. J., 4, *14*, 50, *52*
Delbruck, A., 116, *135*
DeMarsh, O. B., 77, *134*
Dern, R. J., 70, 73, *135*, *137*
Detter, J. C., 75, 77, *135*

DeVillez, E. J., 48, 52
Duke, E., 39

E

Eaton, G. M., 77, 79, *135*, 141, *157*
Emerson, P. M., 140, *157*
Eppenberger, H. M., 108, 110, *135*
Eppenberger, M. E., 108, 110, *135*
Evans, F. T., 88, *135*
Evans, J., 18, *39*

F

Fildes, R. A., 90, 91, *135*
Fine, J., 18, *39*
Fiske, C. H., 129, *135*
Frischer, H., 91, *135*

G

Gall, J. C., 73, 133, 134, *135*, *137*
Gartler, S. M., 141, *157*
Genest, K., 88, *135*
Gershowitz, H., 73, *135*
Giblett, E. R., 75, 77, *135*
Gibson, C. W., 116, *137*
Goldberg, E., 116, *135*
Goldstein, D. P., 94, *136*
Gomori, G., 122, *135*
Gordon, A. H., 4, *14*
Gower, M. K., 91, *135*
Gray, P. W. S., 88, *135*
Green, H., 144, *157*
Green, R. A., 140, *157*
Gregory, K., 140, *157*
Grell, E. H., 118, *135*
Griffith, I. V., 121, *136*
Grossbach, U., 51, *52*
Guttormsen, S. A., 55, *61*

H

Haeffner, L. J., 101, 102, *136*
Hames, C., 73, *135*
Hansl, R., Jr., 51, *52*
Harano, Y., 106, *137*
Harris, H., 6, *14*, 75, 77, 80, 82, 90, 91, 127, 128, 129, *135*, *137*, 150, *157*
Hartman, L., 18, *39*

Hauber, J., 98, *136*
Haupt, I., 116, *137*
Heller, P., 116, *137*
Henderson, N. S., 97, 98, *135*
Hess, A., 122, *135*
Hess, B., 4, *14*
Hill, B. R., 3, *14*, 41, 52
Hirsch, C. A., 98, *135*
Holmes, E. W., Jr., 78, *135*
Honeyman, M. S., 73, 133, 134, *135*, *137*
Hopkinson, D. A., 75, 77, 80, 82, *135*, *137*
Hubby, J. L., 6, *14*, 150, *157*
Hugou, M., 18, *39*
Hume, D. M., 140, *157*
Hunter, R. L., 3, 4, *14*, 86, *136*
Hynick, G., 94, *134*

J

Jacobson, K. B., 118, *135*
Jermyn, M. A., 41, 52
Johnson, F. M., 100, 101, *134*
Jolliff, C., 18, *39*
Jones, R. T., 106, *136*

K

Kalow, W., 88, *135*
Kaplan, J. C., 79, 114, *135*, *136*
Kaplan, N. O., 86, 108, 110, 116, 117, *135*, *136*, 147, *157*
Karcher, D., 43, *52*
Kay, R. H., 164, *175*
Keil, B., 4, *14*
King, E. J., 128, *137*
King, J., 6, *14*
Kitto, G. B., 117, *136*
Klingenburg, M., 116, *135*
Knutsen, C. A., 77, 79, 119, 121, *135*, *136*, 145, 146, 151, *157*
Kobara, T. Y., 112, *136*
Koch, H., 18, *39*
Koen, A. L., 119, *136*, *137*
Kohn, J., 4, *14*
Koler, R. D., 106, *136*
Kottke, M. E., 117, *136*
Kowlessar, O. D., 101, 102, *136*
Krakowski, R. E., 141, *157*
Kunkel, H. G., 4, *14*

L

Lapp, C., 105, 106, *134*
Latner, A. L., 21, 39, 44, 51, 52, 84, 87, 93, 94, 101, 115, 116, 124, *136*, 138, 139, 140, *157*
Lawrence, S. H., 101, *136*
Leavell, B. S., 111, 112, *137*
Lehmann, H., 88, 124, *135*, *136*
Leone, J., 117, 118, *137*
Levine, L., 86, *135*, 147, *157*
Lewontin, R. C., 6, *14*, 150, *157*
Li, C., 6, *14*
Liddell, J., 124, *136*
Lin, E. C. C., 98, *135*
Linder, D., 141, *157*
Locher, J., 140, *157*
Lombroso, L., 49, 52
Long, W. K., 114, *136*
Lowenstein, J. M., 98, *136*
Lowenthal, A., 43, 52
Luck, D. J. L., 117, *136*
Lyon, M., 6, *14*

M

Maas, D. W., 97, 98, *135*
McKusick, V. A., 5, *14*
Macolalag, E. V., 140, *157*
Malone, J. L., 78, *135*
Markert, C. L., 3, 4, 9, *14*, 86, 115, 116, *136* 147, *157*
Marsh, C., 18, 39
Meister, A., 4, *14*
Melnick, P. J., 101, *136*
Meyer, R., 39
Migeon, B. R., 144, *157*
Milkman, R., 6, *14*
Miller, C. S., 144, *157*
Moller, F., 4, 9, *14*, 86, 115, 116, *136*
Monis, B., 101, *136*
Moretti, J., 18, 39
Mori, R., 106, *137*
Morimura, H., 106, *137*
Morrison, M., 73, *136*
Moss, D. W., 128, *137*
Motulsky, A. G., 73, *136*
Mouray, H., 18, 39
Munkres, K. D., 117, *136*
Murphy, J. B., 118, *135*
Murphy, W. H., 117, *136*

N

Neilands, J. B., 3, *14*
Nerenberg, S. T., 38, 39, 45, 46, 47, 51, 52
Nishimura, E. T., 112, *136*
Nisselbaum, J. S., 132, *136*
Nitowsky, H. M., 6, *14*, 73, *135*, 143, *157*

O

Ohno, S., 73, *136*
Oort, J., 43, 52
Ornstein, L., 4, *14*, 50, 52
Ortanderl, F., 4, *15*, 45, 52
Oski, F. A., 78, *135*
Owen, J. A., 133, *136*

P

Paglia, M. D., 77, *134*
Payne, H. W., 73, *136*
Payne, L., 18, 39
Pellegrino, C., 3, *14*, 41, 52
Peng, K. C., 172, *175*
Pfleiderer, G., 3, 4, *15*, 41, 45, 52, 116, *137*
Pierce, J. E., 94, *136*
Pineda, E. P., *136*
Porter, I. H., 5, *14*
Poulik, M., 38, 39
Povey, S., 75, 77, *135*
Prout, G. R., 140, *157*
Pryse-Davies, J., 41, 52
Pun, J. Y., 49, 52

R

Ramsey, H., 18, 39
Rasminsky, M., 98, *135*
Raymond, S., 4, *14*, 51, 52
Reed, T., 6, *14*
Reich, E., 117, *135*, *136*
Ressler, N., 119, *136*
Richterich, R., 140, *157*
Rigas, D. A., 106, *136*
Riggs, S. K., 5, *15*, 89, *137*
Robinson, J. C., 94, *136*
Rosalki, S. B., 45, 52, 108, 110, *136*
Rossi, C., 98, *136*
Rossi, E., 140, *157*

Ruddle, F. H., 117, *137*
Rutenberg, A. M., 45, *52*, 102, *137*

S

Sactor, B., 116, *136*
Sandler, M., 94, *136*
Sanghvi, L., 6, *14*
Sayre, F. W., 3, *14*, 41, *52*
Scandalios, J. G., 101, 102, 103, 112, 117, 118, *136*, 152, *157*
Schoch, H. K., 140, *157*
Schulze, J., 5, *14*
Schwartz, D., 5, *14*
Schwartz, M. K., 132, *136*
Scott, E. M., 121, *136*
Sears, E. R., 94, 95, 97, *135*, 153, 154, 155, *157*
Sebesta, K., 4, *14*
Selagi, S., 144, *157*
Shannon, L. M., 102, 103, *137*, 152, *157*
Shaw, C. R., 5, 10, *14*, 73, 86, 119, *136*, *137*, 147, 148, 149, 150, *157*
Shaw, M. W., 5, *14*, 86, 88, *137*
Shokeir, M., 133, 134, *137*
Shows, T. B., 70, 73, 117, *137*
Shreffler, D. C., 55, *61*, 73, *135*
Silk, E., 88, *135*
Simon, E. R., 77, *134*
Sing, C. F., 55, *61*, 79, 94, 95, 97, 113, 115, *135*, *136*, 145, 146, 151, 153, 154, 155, 156, *157*, 174, *175*
Singer, T. P., 98, *136*
Skillen, A. W., 21, *39*, 44, 51, *52*, 84, 87, 93, 101, 115, 116, 124, *136*, 138, 139, 140, *157*
Slater, R. J., 4, *14*
Sleisenger, M. H., 101, 102, *136*
Smith, E. E., 45, *52*, 102, *137*
Smith, H., 133, *136*
Smith, S. R., 98, *136*
Smithies, O., 4, *14*, 38, *39*, 131, 133, *137*
Snedecor, G. W., 169, 172, *175*
Somers, G. F., 88, *135*
Spencer, H. H., 140, *157*
Spencer, N., 80, 82, 127, 128, 129, *135*, *137*
Starkweather, W. H., 140, 150, *157*

Steel, R. G. D., 169, 172, *175*
Stitzer, K., 119, *136*
Strole, W. B., 111, 112, *137*
Subbarow, Y., 129, *135*
Sur, B. K., 128, *137*
Suter, H., 113, *134*
Sved, J., 6, *14*
Szeinberg, A., 141, *157*

T

Tanaka, T., 106, *136*
Tashian, R. E., 5, *14*, 70, 73, 77, 79, 82, 86, 87, 88, 89, 91, 124, *135*, *137*, 141 150 *157*
Thomas, R., 41, *52*
Thompson, P., 106, *136*
Thorne, C. J. A., 116, *137*
Thorup, O. A., Jr., 111, 112, *137*
Tiselius, A., 3, *15*
Torrie, J. H., 169, 172, *175*
Trujillo, J. M., 105, 106, *134*
Tsao, M. U., 97, 98, 115, 116, *137*

U

Ursprung, H., 117, 118, *137*

V

Valentine, W. N., 77, *134*
Vanbellinghen, P., 106, *136*
Van derHelm, H. J., 124, 125, *137*
Van Sande, M., 43, *52*
Vesell, E. S., 4, *15*, 116, *137*

W

Ways, P. O., 75, 77, *134*, *135*
Weimer, H. E., 101, *136*
Weintraub, L., 4, *14*
Weiss, M. C., 144, *157*
Weitkamp, L. R., 55, *61*
Wieland, T., 3, 4, *15*, 41, 45, *52*, 116, *137*
Wieme, R. J., 4, *15*, 43, *52*
Wilkinson, J. H., 41, 51, *52*, 84, 101, *137*, 138, 140, *157*
Williams, R. A. D., 95, 96, *137*
Willighagen, R. G. J., 43, *52*
Winegrad, A. J., 78, *135*

Worner, W., 116, *137*
Wroblewski, F., 140, *157*

Y

Yakulis, V. J., 116, *137*
Yonetani, T., 84, *137*
Yu, Y. S. L., 5, *15*, 89, *137*

Z

Zebe, E., 116, *135*
Ziprowski, L., 141, *157*
Zuelzer, W., *39*
Zuppinger, J., 140, *157*
Zwilling, E., 86, *135*, 147, *157*

Subject Index

A

Acetylcholinesterase
  isozyme method for, 122–123
Achromatic Regions
  isozyme method for, 82–84
Acid Phosphatase
  isozyme method for, 127–129
Acrylamide Gel Electrophoresis, 47–51
  advantages of, 47–48
  disadvantages of, 48
  slab gel formation, 48–50
  tube gel formation, 50–51
Adenosine Triphosphate Detecting Staining System, 67
Adenylate Kinase
  isozyme method for, 90–92
Agar Gel Electrophoresis, 43–45
Agar Overlay Staining Method, 36
Alcohol Dehydrogenase
  isozyme method for, 117–119
Aldolase
  isozyme method for, 105–106
Alkaline Phosphatase
  isozyme method for, 92–94
Alpha-Glycerophosphate Dehydrogenase
  isozyme method for, 94–95
Amylase
  isozyme method for, 133
Animal Tissues
  sample preparation, 58–59
ATPase
  isozyme method for, 129–131

B

Blood
  anticoagulants, 53–54
  preparation for electrophoresis, 53–54
  storage for electrophoresis, 54–55

Buffers: general principles of use in electrophoresis, 7–8, 63

C

Carbonic Anhydrase
  isozyme method for, 88–90
Catalase
  isozyme method for, 111–113
Cell Cultures
  preparation for electrophoresis, 56–57
Cellulose Acetate Electrophoresis, 45–47
Cellulose Acetate Overlay Staining Method, 36, 37
Ceruloplasmin
  isozyme method for, 133–134
Creatine Kinase
  isozyme method for, 108–110
Current, Electrical
  general principles of use in electrophoresis, 7–8

D

Destaining
  of starch gels, 38
Diaphorase
  isozyme method for, 119–120
Differentiation
  isozymes in the study of, 145–146
DPNH-Methemoglobin Reductase (see Diaphorase)

E

Electroosmosis, 7–8
Electrophoresis
  agar gel, 4
  analytical, 4
  media for, 3–6
  molecular sieving in, 4

Subject Index / 183

moving boundary, 3
paper, 3–4
preparative, 4
starch block, 4
starch gel, 4, 16–39
zone, 3
Electrophoretic Data
    acquisition of, 159–160
    analysis of, 158–175
Electrophoretic Data Collection
    experimental design for, 171–172
    local control, 172–173
    randomization, 172
    replication, 172
Electrophoretic Data Recording
    automated, 161
    densitometry, 161–165
    methods of, 160–165
    storage and handling, 165–168
Electrophoretic Data Reduction, 168–175
    statistics and parameters, 168–169
    variation, 169–170
Electrophoretic Enzyme Systems, 70–134
Electrophoretic Media
    acrylamide gel, 47–51
    agar gel, 43–45
    cellulose acetate, 45–47
    choice of, 51
    paper, 40–43
    starch gel, 17–39
Electrophoretic Setup for Starch Gel, 24–27
    bridge buffer trays, 24, 25
    electrodes, 24, 25
    power supply, 25–27
    stand, 24, 25
    wicks, 24, 25
Electrophoretic Variation (see specific enzyme systems)
    genetically determined, 5–6, 10
Enzymes
    biochemical relationships determined by isozyme techniques, 151–152
    substrate specificities determined by isozyme techniques, 151–152
Erythrocytes
    preparation for electrophoresis, 53–55
Esterase
    isozyme method for, 86–88

Evolution
    studies with isozyme techniques, 153–156

F

Filter Paper Overlay Staining Method, 36, 37
Fructokinase
    isozyme method for, 79–80
Fumarase
    isozyme method for, 104–105

G

Gel Tray Lids, 19–20
Gel Trays, 17–19
Genetic Variation
    isozyme studies to determine extent of, 6
    uses of in isozyme studies, 147–150
Genetics
    developmental, 146–147
    population, 150–151
    somatic cell, 143–145
Glucokinase
    isozyme method for, 77–79
Glucose-6-Phosphate Dehydrogenase
    isozyme method for, 70–73
Glutamate Dehydrogenase
    isozyme method for, 124–125
Glutamic-Oxaloacetic Transaminase
    isozyme method for, 132–133
Glutathione Reductase
    isozyme method for, 113–115
Glyceraldehyde-3-Phosphate Dehydrogenase
    isozyme method for, 95–96
Glycerolization
    of starch gels, 38
Glycerophosphate Dehydrogenase
    isozyme method for, 94–95
Glycolytic Pathways, 66

H

Hemoglobin
    electrophoretic method for, 131–132
Heteropolymers, 11
Hexokinase (see Glucokinase)
Homopolymers, 11

Subject Index

I

Indophenol Oxidase (see Achromatic Regions)
Isocitrate Dehydrogenase
  isozyme method for, 97–98
Isozyme Approach, 1–14
  genetic variation, 5, 6
  history, 3–6
  in the study of ontogenetic variation, 1
  in the study of tissue variation, 1
  principles, 6–8
  simplicity, 1
  specificity, 1
  strengths of, 1–2
Isozyme Bands
  numbering of, 12–14
Isozyme Techniques, 70–134
  acetylcholinesterase, 122–123
  achromatic regions, 82–84
  acid phosphatase, 127–129
  adenylate kinase, 90–92
  alcohol dehydrogenase ADH, 117–119
  aldolase, 105–106
  alkaline phosphatase, 92–94
  α-glycerophosphate dehydrogenase (α-GPD), 94–95
  amylase, 133
  ATPase, 129–131
  carbonic anhydrase (CA), 88–90
  catalase, 111–113
  ceruloplasmin, 133–134
  creatine kinase (CK), 108–110
  diaphorase, 119–121
  DPNH-methemoglobin reductase (see diaphorase)
  esterase, 86–88
  for determining substrate specificities of enzymes, 151–152
  fructokinase, 79–80
  fumarase, 104–105
  general protein staining method, 132
  glucokinase, 77–79
  glucose-6-phosphate dehydrogenase (G-6-PD), 70–73
  glutamate dehydrogenase, 124–125
  glutamic-oxaloacetic transaminase (GOT), 132–133
  glutathione reductase (GR), 113–115
  glyceraldehyde-3-phosphate dehydrogenase (GA-3-PD), 95–96
  hemoglobin identification, 131–132
  hexokinase (see glucokinase)
  history of, 2–6
  indophenol oxidase activity (see achromatic regions)
  in tissue culture, 143–145
  isocitrate dehydrogenase (ICD), 97–98
  lactate dehydrogenase (LDH), 84–86
  leucine aminopeptidase (LAP), 100–102
  malate dehydrogenase (MDH), 115–117
  peroxidase, 102–104
  phosphoglucomutase (PGM), 80–82
  6-phosphogluconate dehydrogenase (6-PGD), 73–75
  phosphohexose isomerase (PHI), 75–77
  pseudocholinesterase, 123–124
  pyruvate kinase (PK), 106–108
  succinate dehydrogenase (SDH), 98–99
  triosephosphate isomerase (TPI), 121–122
  xanthine dehydrogenase (XDH), 126–127
Isozymes
  conformational, 11
  definition of, 8–10
  in plants, 152–153
  in the study of evolution, 153–156
  isokinetic, 11
  nomenclature, 8–14
  nonsegregatnig, 10–11
  terminology, 8–14
  types of, 10–12
Isozymology
  clinical applications, 138–145
  clinical applications of serum isozymes, 138–140
  clinical applications in neoplastic disease, 140–141
  future clinical applications of, 141–143
  in somatic cell genetics, 143–145
  in study of developmental genetics, 146–147

in studies of genetic variation, 147–150
in the study of population genetics, 150–151
in the study of tissue-organ and intracellular differentiation, 145–146

L

Lactate Dehydrogenase
  isozyme method for, 84–86
Leucine Aminopeptidase
  isozyme method for, 100–102
Leukocytes and Platelets
  preparation for electrophoresis, 55–56

M

Malate Dehydrogenase
  isozyme method for, 115–117
Microorganisms
  preparation for electrophoresis, 61
Molecular Sieving
  principle of, 7

N

Neoplastic Disease
  application of isozymology in, 140–141

P

Paper Electrophoresis, 40–43
Peroxidase
  isozyme method for, 102–104
Phosphoglucomutase
  isozyme method for, 80–82
6-Phosphogluconate Dehydrogenase
  isozyme method for, 73–75
Phosphohexose Isomerase
  isozyme method for, 75–77
Plant Isozymes, 152–153
Plant Tissues
  preparation for electrophoresis, 60–61
Platelets and Leukocytes
  sample preparation, 55–56
Power Supply
  for starch gel electrophoresis, 25
Protein Staining Method, 132
Pseudocholinesterase
  isozyme method for, 123–124
Pyruvate Kinase
  isozyme method for, 106–108

R

Reagents for Electrophoresis, 67–70

S

Sample Addition Technique, 23–24
Sample Preparation for Electrophoresis
  animal soft tissues, 58–59
  cell cultures, 56–57
  erythrocytes, 53–55
  leukocytes and platelets, 55–56
  microorganisms, 61
  plant tissues, 60–61
  serum, 57–58
  wheat seeds, 60
Serum
  preparation for electrophoresis, 57–58
Serum Isozymes
  applied to clinical medicine, 138–140
Slicers
  for starch gels, 32–34
Staining of Starch Gels, 34–37
  agar overlay method, 36
  cellulose acetate overlay method, 36–37
  filter paper overlay method, 36–37
  standard solution method, 34–36
Staining Systems
  adenosine triphosphate detecting, 67
  general, 63–67
  tetrazolium, 67
Starch Gel
  cooling of, 27–29
  destaining of, 38
  glycerolization of, 38
  preparation of, 21–23
    by short method, 23
    by standard method, 21–23
  preservation of, 37
  recording data from, 37–38
  sample addition to, 23–24
  setting up of, 29–30
  slicing of, 31–34
  special studies with, 38–39
  staining of, 34–37
    by agar overlay method, 36
    by cellulose acetate or filter paper overlay method, 36–37
    by standard solution method, 34–36

Starch Gel Electrophoresis, 16–39
  duration of, 31
  power supplies for, 25–27
Starch Gel Equipment, 17–25, 33–34
  bridge buffer trays, 24, 25
  electrodes, 24, 25
  slicers, 32–34
  stands, 24–25
  templates, 20–21
  tray lids, 19–20
  trays, 17–19
  wicks, 24, 25
Statistics
  in electrophoretic data reduction, 168–169
Succinate Dehydrogenase
  isozyme method for, 98–99

T

Templates, 20–21
Tetrazolium Staining System, 67

Tissue, Animal
  preparation for electrophoresis, 58–59
  storage for electrophoresis, 58–59
Tissue Culture
  isozyme techniques in, 143–145
Tissues, Plant
  preparation for electrophoresis, 60–61
Triosephosphate Isomerase
  isozyme method for, 121–122

W

Wheat Seeds
  preparation for electrophoresis, 60

X

Xanthine Dehydrogenase
  isozyme method for, 126–127
X-chromosomal Inactivation
  study with isozymes, 6